Animal Science

Animal Science

Vera Shaffer

Larsen & Keller
www.larsen-keller.com

Animal Science
Vera Shaffer
ISBN: 978-1-64172-148-6 (Hardback)

Larsen & Keller
Published by Larsen and Keller Education,
5 Penn Plaza,
19th Floor,
New York, NY 10001, USA

Cataloging-in-Publication Data

Animal science / Vera Shaffer.
p. cm.
Includes bibliographical references and index.
ISBN 978-1-64172-148-6
1. Zoology. 2. Animals. I. Shaffer, Vera.

QL45.2 .A55 2019
590--dc23

Printed and bound in China.

For more information regarding Larsen and Keller Education and its products, please visit the publisher's website www.larsen-keller.com

Table of Contents

Preface

Animal science is the study of the biology of animals, which live under human control. These animals include livestock species as well as pets. The fundamental themes of study in animal science are the areas of animal breeding, food and fiber production, nutrition as well as the understanding of animal behavior and welfare. The discipline of animal science is approached from a scientific framework of principles of genetics, microbiology, reproduction and physiology. This book attempts to understand the multiple branches that fall under the discipline of animal science and how such concepts have practical applications. While understanding the long-term perspectives of the topics, the textbook makes an effort in highlighting their impact as a modern tool for the growth of the field. It will serve as a reference to a broad spectrum of readers.

A short introduction to every chapter is written below to provide an overview of the content of the book:

Chapter 1- Animal science is a branch of biology, which is concerned with the study of animals. This encompasses investigation into the structure, embryology, classification, evolution, habits and distribution of animals as well as their interaction with their environment. This is an introductory chapter, which will introduce briefly all the significant topics related to the study of animal and animal kingdom, as well as animal behavior, anatomy and physiology, genetics, etc.; **Chapter 2-** Vertebrates are animals that have backbones, which are surrounded by cartilage or bone. Invertebrates do not have a spinal cord. The topics elaborated in this chapter will help in providing a better perspective about vertebrate and invertebrate animals. It elucidates the five primary vertebrate groups and seven chief invertebrate groups; **Chapter 3-** Animals can be categorized according to their feeding habits into the ecological groups of herbivores, carnivores, omnivores, fungivores and bacterivores. This chapter closely examines these key animal groups in great detail to provide an extensive understanding of their feeding behavior; **Chapter 4-** The study of mental cognition or intelligence involves a study of the different mental capacities of animals. Some of the animals whose intelligence is widely studied are dogs, cats, pigeons, elephants, etc. This chapter presents an overview of intelligence in such animals and also discusses significant topics, such as the theory of mind in animals and the mechanism of animal navigation.

Finally, I would like to thank my fellow scholars who gave constructive feedback and my family members who supported me at every step.

Vera Shaffer

1

Introduction to Animal Science

Animal science is a branch of biology, which is concerned with the study of animals. This encompasses investigation into the structure, embryology, classification, evolution, habits and distribution of animals as well as their interaction with their environment. This is an introductory chapter, which will introduce briefly all the significant topics related to the study of animal and animal kingdom, as well as animal behavior, anatomy and physiology, genetics, etc.

Animal

Animal, (kingdom Animalia), any of a group of multicellular eukaryotic organisms (*i.e.,* as distinct from bacteria, their deoxyribonucleic acid, or DNA, is contained in a membrane-bound nucleus). They are thought to have evolved independently from the unicellular eukaryotes. Animals differ from members of the two other kingdoms of multicellular eukaryotes, the plants (Plantae) and the fungi (Mycota), in fundamental variations in morphology and physiology. This is largely because animals have developed muscles and hence mobility, a characteristic that has stimulated the further development of tissues and organ systems.

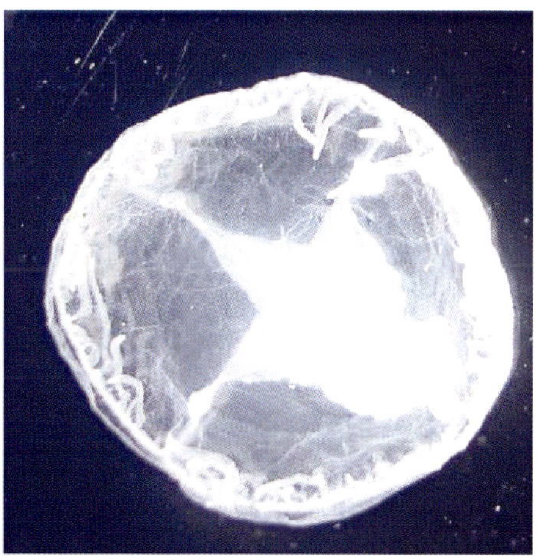

Freshwater jellyfish (*Craspedacusta sowerbyi*).*U. S. Geological Survey*

Animals dominate human conceptions of life on Earth not simply by their size, abundance, and sheer diversity but also by their mobility, a trait that humans share. So integral is movement to the conception of animals that sponges, which lack muscle tissues, were long considered to be plants. Only after their small movements were noticed in 1765 did the animal nature of sponges slowly come to be recognized.

In size animals are outdone on land by plants, among whose foliage they may often hide. In contrast, the photosynthetic algae, which feed the open oceans, are usually too small to be seen, but marine animals range to the size of whales. Diversity of form, in contrast to size, only impinges peripherally on human awareness of life and thus is less noticed. Nevertheless, animals represent three-quarters or more of the species on Earth, a diversity that reflects the flexibility in feeding, defense, and reproduction which mobility gives them. Animals follow virtually every known mode of living that has been described for the creatures of Earth.

Gray whale (*Eschrichtius robustus*) breaching.

Animals move in pursuit of food, mates, or refuge from predators, and this movement attracts attention and interest, particularly as it becomes apparent that the behaviour of some creatures is not so very different from human behaviour. Other than out of simple curiosity, humans study animals to learn about themselves, who are a very recent product of the evolution of animals.

Giraffa camelopardalis by David Bygott

It is estimated that around 9 or 10 million species of animals inhabit the earth; the exact number is not known and all estimates are rough. Animals range in size from no more than a few cells to organisms weighing many tons, such as the giant squid. By far most species of animals are insects, with groups such as mollusks, crustaceans, and nematodes also being especially diverse. By this measure our own group, the vertebrates, is relatively inconsequential from a diversity perspective.

Characteristics

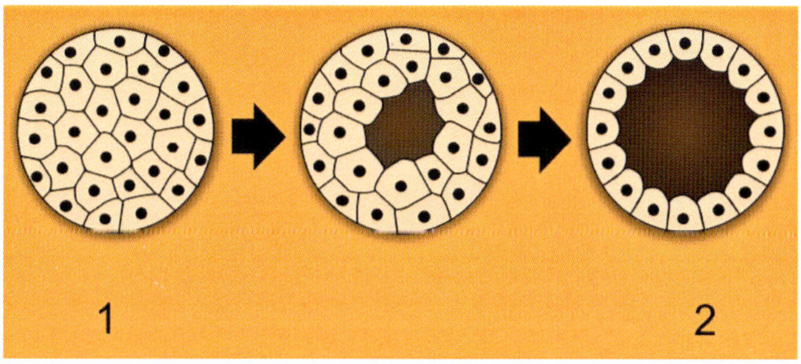

Animals are unique in having the ball of cells of the early embryo (1) develop into a hollow ball or blastula (2).

Animals have several characteristics that set them apart from other living things. Animals are eukaryotic and multicellular, unlike bacteria, which are prokaryotic, and unlike protists, which are eukaryotic but unicellular. Unlike plants and algae, which produce their own nutrients animals are heterotrophic, feeding on organic material and digesting it internally. With very few exceptions, animals breathe oxygen and respire aerobically. All animals are motile (able to spontaneously move their bodies) during at least part of their life cycle, but some animals, such as sponges, corals, mussels, and barnacles, later become sessile. The blastula is a stage in embryonic development that is unique to most animals, allowing cells to be differentiated into specialised tissues and organs.

Structure

All animals are composed of cells, surrounded by a characteristic extracellular matrix composed of collagen and elastic glycoproteins. During development, the animal extracellular matrix forms a relatively flexible framework upon which cells can move about and be reorganised, making the formation of complex structures possible. This may be calcified, forming structures such as shells, bones, and spicules. In contrast, the cells of other multicellular organisms (primarily algae, plants, and fungi) are held in place by cell walls, and so develop by progressive growth. Animal cells uniquely possess the cell junctions called tight junctions, gap junctions, and desmosomes.

With few exceptions—in particular, the sponges and placozoans—animal bodies are differentiated into tissues. These include muscles, which enable locomotion, and nerve tissues, which transmit signals and coordinate the body. Typically, there is also an internal digestive chamber with either one opening (as in flatworms) or two openings (as in deuterostomes).

Reproduction and Development

Sexual reproduction is nearly universal in animals, such as these dragonflies.

Nearly all animals make use of some form of sexual reproduction. They produce haploid gametes by meiosis; the smaller, motile gametes are spermatozoa and the larger, non-motile gametes are ova. These fuse to form zygotes, which develop via mitosis into a hollow sphere, called a blastula. In sponges, blastula larvae swim to a new location, attach to the seabed, and develop into a new sponge. In most other groups, the blastula undergoes more complicated rearrangement. It first invaginates to form a gastrula with a digestive chamber and two separate germ layers, an external ectoderm and an internal endoderm. In most cases, a third germ layer, the mesoderm, also develops between them. These germ layers then differentiate to form tissues and organs.

Repeated instances of mating with a close relative during sexual reproduction generally leads to inbreeding depression within a population due to the increased prevalence of harmful recessive traits. Animals have evolved numerous mechanisms for avoiding close inbreeding. In some species, such as the splendid fairywren (*Malurus splendens*), females benefit by mating with multiple males, thus producing more offspring of higher genetic quality.

Some animals are capable of asexual reproduction, which often results in a genetic clone of the parent. This may take place through fragmentation; budding, such as in *Hydra* and other cnidarians; or parthenogenesis, where fertile eggs are produced without mating, such as in aphids.

Animal Kingdom

Animals evolved from unicellular eukaryotes. The presence of a nuclear membrane in eukaryotes permits separation of the two phases of proteinsynthesis: transcription (copying) of deoxyribonucleic acid (DNA) in the nucleus and translation (decoding) of the message into protein in the cytoplasm. Compared to the structure of the bacterial cell, this gives greater control over which proteins are produced. Such control permits specialization of cells, each with identical DNA but with the ability to control finely which genes successfully send copies into the cytoplasm. Tissues and organs can thus evolve. The semirigid cell walls found in plants and fungi, which constrain the shape and hence the diversity of possible cell types, are absent in animals. If they were present, nerve and muscle cells, the focal point of animal mobility, would not be possible.

All animals are members of the Kingdom Animalia, also called Metazoa. This Kingdom does not contain prokaryotes (organisms made up of cells that lack a cell nucleus or any membrane-encased organelles, such as bacteria and blue-green algae) or protists (Kingdom Protista, which includes unicellular eukaryotic organisms). All members of Animalia are multicellular, and all are heterotrophs (i.e., they rely directly or indirectly on other organisms for their nourishment). Most ingest food and digest it in an internal cavity.

A characteristic of members of the animal kingdom is the presence of muscles and the mobility they afford. Mobility is an important influence on how an organism obtains nutrients for growth and reproduction. Animals typically move, in one way or another, to feed on other living organisms, but some consume dead organic matter or even photosynthesize by housing symbiotic algae. The type of nutrition is not as decisive as the type of mobility in distinguishing animals from the other two multicellular kingdoms. Some plants and fungi preyon animals by using movements based on changing turgor pressure in key cells, as compared with the myofilament-based mobility seen in animals. Mobility requires the development of vastly more elaborate senses and internal communication than are found in plants or fungi. It also requires a different mode of growth: animals increase in size mostly by expanding all parts of the body, whereas plants and fungi mostly extend their terminal edges.

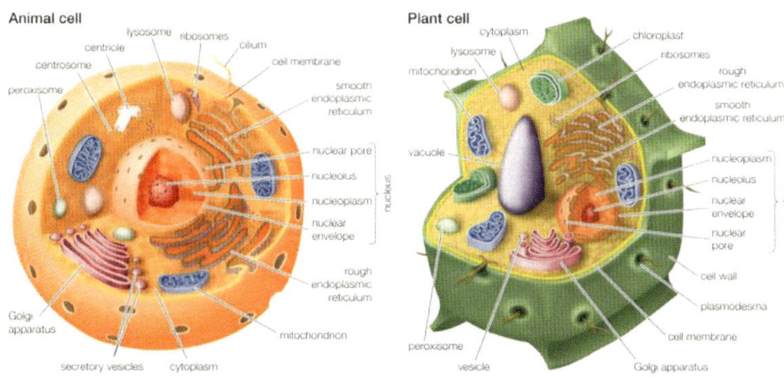

Cytoplasm is contained within cells in the space between the cell membrane and the nuclear membrane.

All phyla of the animal kingdom, including sponges, possess collagen, a triple helix of protein that binds cells into tissues. The walled cells of plants and fungi are held together by other molecules, such as pectin. Because collagen is not found among unicellular eukaryotes, even those forming colonies, it is one of the indications that animals arose once from a common unicellular ancestor.

The muscles that distinguish animals from plants or fungi are specializations of the actin and myosin microfilaments common to all eukaryotic cells. Ancestral sponges, in fact, are in some ways not much more complex than aggregations of protozoans that feed in much the same way. Although the sensory and nervous system of animals is also made of modified cells of a type lacking in plants and fungi, the basic mechanism of communication is but a specialization of a chemical system that is found in protists, plants, and fungi. The lines that divide an evolutionary continuum are rarely sharp.

Mobility constrains an animal to maintain more or less the same shape throughout its active life. With growth, each organ system tends to increase roughly proportionately. In contrast, plants and fungi grow by extension of their outer surfaces, and thus their shape is ever changing. This basic difference in growth patterns has some interesting consequences. For example, animals can rarely sacrifice parts of their bodies to satisfy the appetites of predators (tails and limbs are occasionally exceptions), whereas plants and fungi do so almost universally.

The Animal Kingdom has by far the most number of species among all kingdoms. In fact, you could add up all the known species of the other four kingdoms and this figure would still be less than those found in the Animal Kingdom. Although the Animal Kingdom is highly diverse, all organisms in this kingdom share three main characteristics: they rely on other animals and plants for nourishment, they are mobile, and are multicellular.

Most animals reproduce sexually and have sense organs that allow them to hear, touch, see, taste, and smell. Many are made of millions of cells. These millions of cells can divide into tissue based on their function. When there are many tissues working together, they form an organ. When the organ works with other organs on a common task, this is known as an organ system. These levels of organization are unique to the Animal Kingdom.

Because the Animal Kingdom is so large, it is subdivided into other categories which are then also further divided. These subdivisions are based on animals' similar characteristics or lack thereof.

Animal Behavior

The study of animal behavior begins with understanding how an animal's physiology and anatomy are integrated with its behavior. Both external and internal stimuli prompt behaviors – external information (e.g., threats from other animals, sounds, smells) or weather and internal information (e.g., hunger, fear). Understanding how genes and the environment come together to shape animal behavior is also an important underpinning of the field. Genes capture the evolutionary responses of prior populations to selection on behavior. Environmental flexibility gives animals the opportunity to adjust to changes during their own lifetime.

Behavior can also be defined more narrowly as a change in the activity of an organism in response to a *stimulus*, an external or internal cue or combination of cues.

For example, your dog might start drooling—a change in activity—in response to the sight of food—a stimulus.

Behavioral biology is the study of the biological and evolutionary bases for behavior. Modern behavioral biology draws on work from the related but distinct disciplines of ethology and comparative psychology.

- Ethology is a field of basic biology, like ecology or genetics. It focuses on the behaviors of diverse organisms in their natural environment.
- Comparative psychology is an extension of work done in human psychology. It focuses largely on a few species studied in a lab setting.

Behavioral biology also draws on many related areas of biology, including genetics, anatomy, physiology, evolutionary biology, and, of course, neurobiology—which traces the neural circuits that underlie animal behavior.

Four Questions to Understand a Behavior

Nikolaas (Niko) Tinbergen was a Dutch ornithologist, or bird biologist, who studied behavior and is now considered one of the founders of the field of ethology. Based on his own research, Tinbergen proposed four basic questions helpful in understanding any animal behavior.

Let's look at these questions, using the production of song by the zebra finch.

1. Causation—What causes the behavior? What triggers the behavior, and what body parts, functions, and molecules are involved in carrying it out?

   Example: Singing is triggered in zebra finches by social cues, such as the proximity of a potential mate, as well as the appropriate hormonal state. The ability to produce songs is influenced by male hormones and occurs mainly in male birds. Songs are produced when air flows from air sacs in the bronchii through an organ called the syrinx. Certain parts of the brain control song production and are well-developed in male zebra finches.

2. Development—How does the behavior develop? Is the behavior present early in life? Does it change over the course of the organism's lifetime? What experiences are necessary for its development?

   Example: Young male zebra finches first listen to the songs of nearby males of their species, particularly their fathers. Then, they start to practice singing. By adulthood, male zebra finches have learned to produce their own songs, which are unique but often have similarities to those of their fathers. Once a finch has perfected its song, the song remains fixed for life.

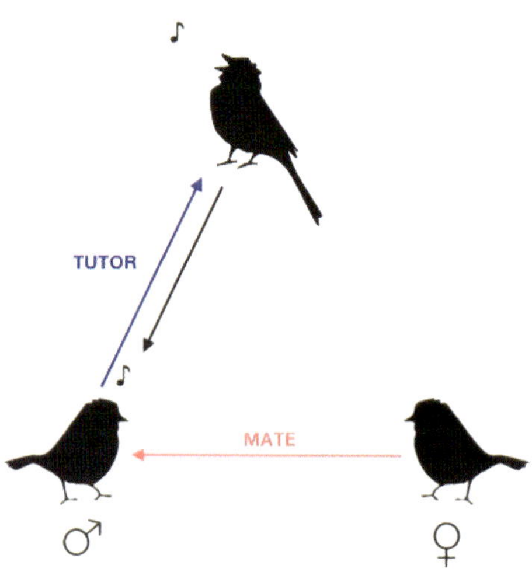

3. Function/adaptive value—How does the behavior affect fitness? How does the behavior affect an organism's chances of survival and reproduction?

   Example: Singing helps male zebra finches attract mates, increasing the chances that they will reproduce. Singing is part of an elaborate courtship ritual that entices the female to choose the male.

4. Phylogeny—How did the behavior evolve? How does the behavior compare to those of related species? Why might it have evolved as it did?

   Example: Almost all species of birds can make vocal sounds, but only those in the suborder *Passeri* are songbirds. Relative to the zebra finch, other songbird species differ in the timing of their listening and practicing phases, the plasticity of song over their lifetimes, the extent to which the song is similar among individuals of the species, and the way that singing is used—for example, for defense of territory vs. courtship of mates.

Cues that Trigger Behavior

At its core, an animal behavior is a response to an internal or external cue. Through behavior, animals can act on the information they receive in ways that will, hopefully, favor their survival and reproductive success.

What kinds of cues can trigger behavior? In some cases, the cue is largely external:

- In hibernation, an animal goes into a den or burrow, reduces its metabolic rate, and enters a state of inactivity during the winter, conserving resources while conditions are harsh and food is scarce. Environmental cues often trigger hibernation behavior. For instance, brown bears enter their den and hibernate when temperature drops and snowfall begins.
- Estivation is similar to hibernation, but it occurs during the summer months. Some desert animals estivate in response to dry conditions. This shift helps them survive the harshest months of the year. The snails in the photo below climb to the tops of fence posts to estivate.

- Migration is a behavior in which animals move from one location to another in a seasonal pattern. For instance, monarch butterflies living in the northern and central United States migrate to Mexico in the autumn, where they spend the winter. Environmental cues that trigger the autumn migration include air temperature, day length, and food availability.

In other cases, the cue for a behavior may be internal. For instance, some behaviors occur with a circadian rhythm, meaning that they are triggered by the animal's internal body clock. You, for example, tend to wake up and become active at roughly the same time each day. As you may have discovered if you've ever taken a long flight, your body's alarm clock will still "go off" at the same time even if the external cues change, which is what causes jet lag.

It's also common for behaviors to be triggered by a combination of internal and external cues interacting. For instance, mating behaviors may be triggered in an animal only when it's in the right hormonal state, and when it sees a member of the opposite sex.

Innate vs. Learned Behaviors

When we are trying to understand how a behavior develops and how it arose evolutionarily, one important question is whether the behavior is genetically preprogrammed or acquired through experience. Let's consider some vocab:

- *Innate* behaviors are genetically hardwired and are inherited by an organism from its parents.
- *Learned* behaviors are not inherited. They develop during an organism's lifetime as the result of experience and environmental influence.

Behavioral biologists have found that many behaviors have both an innate and a learned component. So, it's generally most accurate for us to ask *to what extent* a behavior is innate or learned.

Mostly Innate Behaviors

There are some examples of behaviors that are really and truly hardwired. These behaviors take place in a highly predictable way in response to the right stimulus, even if the organism has never before encountered that stimulus.

For example, an adult salamander will swim perfectly if it's placed in water, even if it never saw water when it was young and has never watched another salamander swim. In this case, the behavior of swimming can only be explained as something genetically preprogrammed in the salamander.

Similarly, you—or any human—will rapidly jerk your hand away if you touch a very hot object. This response is a reflex that's hardwired in the circuits of your sensory and motor neurons and doesn't even involve your brain.

Partly Innate, Partly Learned Behaviors

In other cases, an organism is genetically programmed to develop a behavior, but the form the behavior takes depends on the individual's experience.

One example is the learning of a song by a zebra finch or other songbird, as we saw above. All male zebra finches will begin listening to and learning song at about the same age and practicing and producing song at a slightly later age. Although this pattern is genetically determined, the exact features of the song a bird sings will depend on the songs it hears during its learning period.

Another, more familiar example is language acquisition in humans. Babies are pre-programmed for language learning, but which language they learn depends on what they're exposed to during their plastic, or formative, period.

Mostly Learned Behaviors

In other cases, behaviors are largely dependent on experience—they're learned—and can't be fully explained by genetic preprogramming.

For instance, if a rat receives a food reward each time it pushes a lever, it will quickly learn to push the lever in order to get the food. Similarly, if a cow gets an electric shock each time it brushes up against an electric fence, like the one below, it will rapidly learn to avoid the fence.6^66start superscript, 6, end superscript Pushing a lever to get a reward and avoiding electric fences are not hardwired in rats and cows but are, instead, learned behaviors the animals develop through experience.

If a behavior is learned rather than innate, it isn't directly inherited. But it does still depend on genes. For instance, not all types of animals could learn to push a lever to get a reward. The rat's capacity to learn this behavior depends on how its brain is wired, and the construction, maintenance, and function of a rat brain are all determined by genes in the rat genome.

Animal Anatomy and Physiology

The study of the structure of living things is called anatomy. All animals are made up of CELLS, some of which are specialized to carry out different functions. Simple animals, such as sponges, are made up of only a few types of cell. In more complex animals, cells are organized into tissues, such as muscles and nerves that are necessary for movement. Tissues can form organs, such as the heart, which is used to pump blood around the Circulatory System.

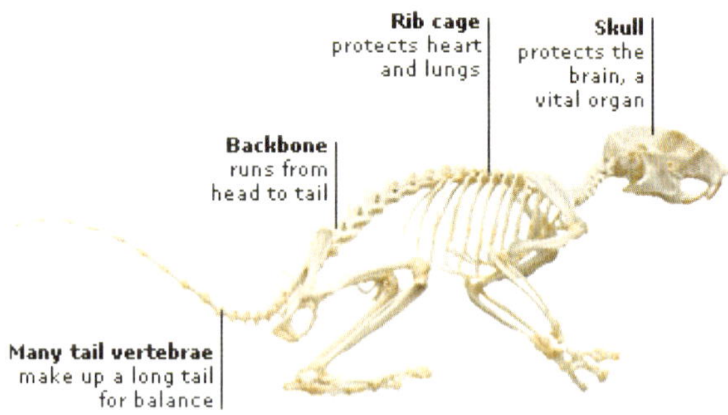

All animals with backbones have an internal framework of support, called an endoskeleton. Bony skeletons, such as that of the squirrel, are light to aid movement. When an animal is young, bones in the skeleton can grow in length. Some bones protect vital organs, while limb bones provide anchorage for muscles.

All animals need oxygen to survive. Simple animals exchange gases over the surface of their bodies. Insects, such as caterpillars, have openings along their bodies, called spiracles, through which air passes. Animals with lungs, such as birds and mammals, actively breathe.

Perfect Symmetry

Most animals are bilaterally symmetrical. If a penguin were cut in half from head to toe, the two halves would be mirror images of each other. Other animals, such as sea anemones, are radially symmetrical. They have no head or tail and can be cut into identical halves along many lines. Of the two types, animals that are bilaterally symmetrical tend to move more quickly and precisely.

Shark Anatomy

Like all fish, sharks have a backbone, breathe through gills, manoeuvre using fins, and are ectothermic (cold-blooded). A shark's anatomy also bears the hallmarks of a predatory fish. They have a streamlined, torpedo-shaped body that allows them to cut easily through water to chase prey. They also have powerful jaws and sharp teeth.

Exoskeleton

Like all arthropods, lobsters have a hard outer casing, called an exoskeleton, made up of plates formed from a substance called chitin. The plates meet at flexible points, such as the leg joints. This exoskeleton provides anchorage for muscles and protection from predators. It also provides support for movement on land and prevents excess water loss.

Cells

Animal cells are typically just 0.02 mm (1/1,250 in) across. Although they can be extremely varied, they share common features. Cells are surrounded by a skin called a membrane and contain a jelly-like fluid called cytoplasm. All the processes needed for life, such as producing energy from food, removing waste, and growth take place inside cells.

Cell Components

Inside an animal cell, the cytoplasm contains structures called organelles that have a variety of functions, from storing vital substances to destroying bacteria. The most important organelle is called the nucleus, which carries genetic information, controlling how the cell behaves. Another organelle, the mitochondrion, produces energy from food.

Circulatory System

The circulatory system carries blood around an animal's body, providing nourishment and oxygen to cells. In some animals it is open, in others it is closed. In an open system, blood flows freely around the body. In a closed system, blood is confined to a network of vessels. The circulatory system also helps distribute heat around the body.

Animal Physiology is the study of how animals work and the various physical and chemical processes that occur within the animal body. It is the key discipline for understanding the fundamental biology of all animals. The various processes of animal body studied under physiology include gas exchange, osmoregulation, blood and circulation, digestion system, nervous system and endocrinology.

Animal Skeleton

Movement is one of the essential features of living things. Cellular movement is observed in one-celled amoebas, ciliates, and flagellates. Flagella whip about to produce cellular motion, while cilia beat synchronously to propel a cell.

In animals, movement is essential for locating food, escaping predators, and seeking mates. In many animals, the movement process is centered in the muscle cell, which contracts and relaxes. The contraction yields great force, which is applied against a surface by means of a skeleton.

Skeletal systems provide structure and protection for a variety of organisms. A water-based skeleton provides the structure necessary for movement in worms. The hard external skeleton (exoskeleton) not only provides a protective mechanism for many organisms but also assists in movement of insects. The internal skeleton (endoskeleton) present in many animals provides the structural network for support, protection, and movement.

Types of Skeletons

Hydrostatic Skeleton

Many animals have a water-based skeleton, or hydrostatic skeleton. Hydrostatic skeletons do not contain hard structures, such as bone, for muscles to pull against. Rather, the muscles surround a fluid-filled body cavity. In a worm, for example, movement occurs when muscle cells contract, and the contractions squeeze internal fluid (the hydrostatic skeleton) against the skin, causing the worm to stiffen and the body to shorten and widen. The squirming motion of a worm also depends on a hydrostatic skeleton.

Exoskeleton

A second type of skeleton, the exoskeleton, exists in arthropods and mollusks. In mollusks, the exoskeleton is a hard, protective outer covering. An example is a clamshell; when a clam's muscles contract, they close the shell rapidly, creating a spurt of water that propels the clam. In arthropods, the exoskeleton also provides protection and movement. Usually, wings are attached by muscles in the hard body surface, which provides a foundation for the muscle contractions. Muscle contractions raise and lower the wings, allowing flight.

Endoskeleton

Vertebrates have an internal skeleton called the endoskeleton, a framework of bones and cartilage that serves as a point of attachment for muscle. The endoskeleton thus transmits the force of muscle contractions. The endoskeleton also provides support for the body (for example, the legs) and protection (the skull).

Bone contains concentric rings of tissue in which bone cells called *osteoblasts* produce the inorganic materials (fibers and matrix) of bone. Much of this material is calcium phosphate, formed from calcium and phosphorus delivered by the blood. Living, mature bone cells called *osteocytes* are also located in the bone. Bone-destroying cells called *osteoclasts* break down bone, thus providing a turnover of bone material needed in other areas. The combination of bone cells and bone tissue comprises a unit called a *Haversian system*. Blood vessels and nerves also exist within the Haversian system.

Bones come together to form a joint, which may be immovable, such as in the sutures of the skull, or movable, such as in the joints of the elbow and shoulder. In a movable joint, a capsule of synovial fluid provides lubrication. Tough, fibrous tissues, known as ligaments, link bones to one another. Connective tissues, called tendons, attach muscles to bones.

The Vertebral Column

The vertebral column consists of a series of bones called vertebrae linked together to form a flexible column with the skull at one end and the tail at the other. Each vertebra consists of a ring of bone with spines (spinous process) protruding dorsally from it. The spinal cord passes through the hole in the middle and muscles attach to the spines making movement of the body possible.

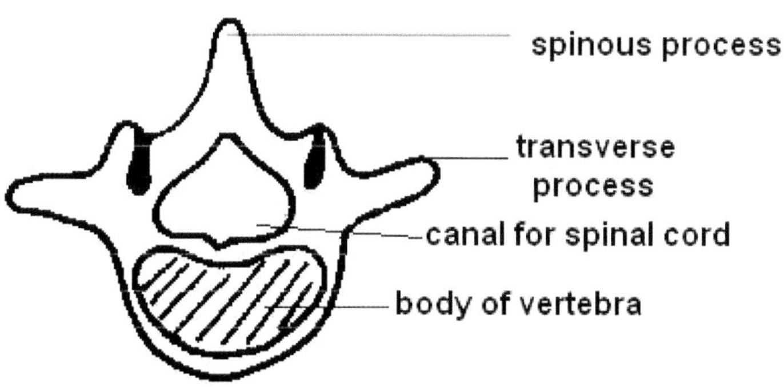

Cross section of a lumbar vertebre

The shape and size of the vertebrae of mammals vary from the neck to the tail. In the neck there are cervical vertebrae with the two top ones, the atlas and axis, being specialized to support the head and allow it to nod "Yes" and shake "No". Thoracic vertebrae in the chest region have special surfaces against which the ribs move during breathing. Grazing animals like cows and giraffes that have to support weighty heads on long necks have extra large spines on their cervical and thoracic vertebrae for muscles to attach to. Lumbar vertebrae in the loin region are usually large strong vertebrae with prominent spines for the attachment of the large muscles of the lower back. The sacral vertebrae are usually fused into one solid bone called the sacrum that sits within the pelvic girdle. Finally there are a variable number of small bones in the tail called the coccygeal vertebrae .

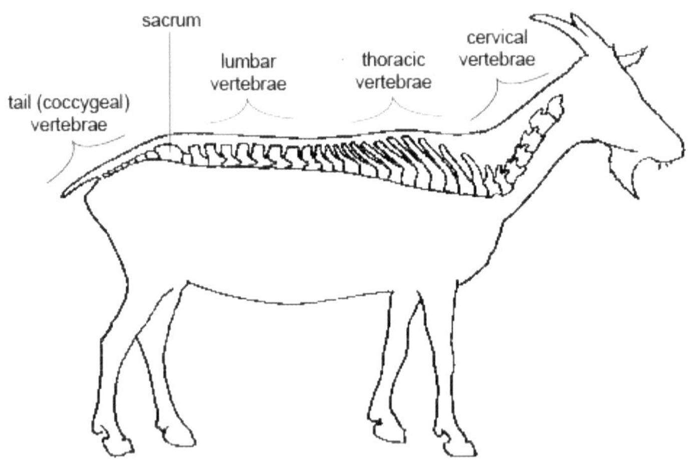

The regions of the vertebral column dik

The Skull

The skull of mammals consists of 30 separate bones that grow together during development to form a solid case protecting the brain and sense organs. The "box "enclosing and protecting the brain is called the cranium . The bony wall of the cranium encloses the middle and inner ears, protects the organs of smell in the nasal cavity and the eyes in sockets known as orbits. The teeth are inserted into the upper and lower jaws (see Chapter 5 for more on teeth) The lower jaw is known as the mandible. It forms a joint with the skull moved by strong muscles that allow an animal to chew. At the front of the skull is the nasal cavity, separated from the mouth by a plate of bone called the palate. Behind the nasal cavity and connecting with it are the sinuses. These are air spaces in the bones of the skull which help keep the skull as light as possible. At the base of the cranium is the foramen magnum, translated as "big hole", through which the spinal cord passes. On either side of this are two small, smooth rounded knobs or condyles that articulate (move against) the first or Atlas vertebra.

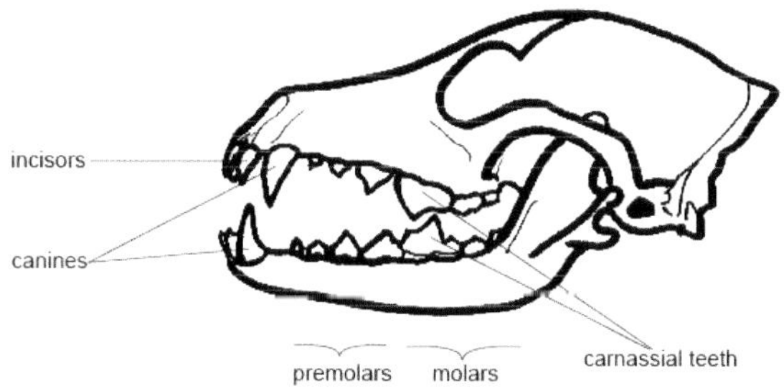

A dog's skull

The Rib

Paired ribs are attached to each thoracic vertebra against which they move in breathing. Each rib is attached ventrally either to the sternum or to the rib in front by cartilage to form the rib cage that protects the heart and lungs. In dogs one pair of ribs is not attached ventrally at all. They are called floating ribs . Birds have a large expanded sternum called the keel to which the flight muscles (the 'breast" meat of a roast chicken) are attached.

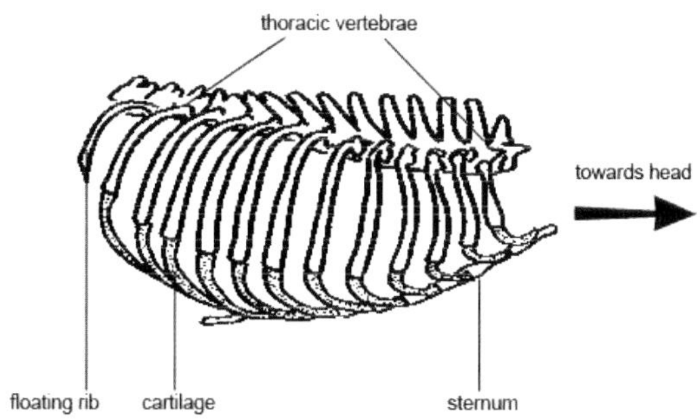

The rib

The Forelimb

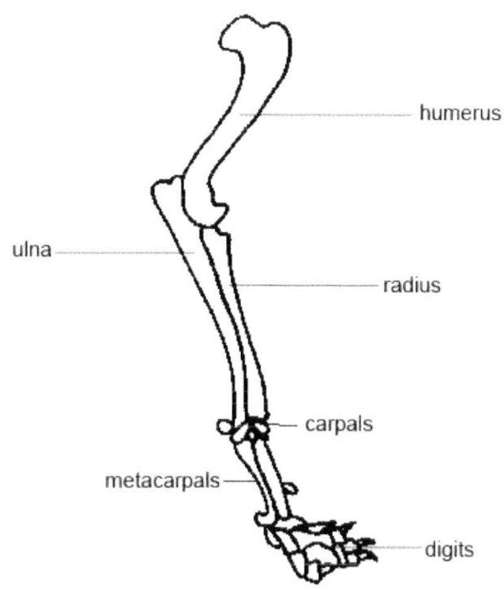

Forelimb of a dog

The forelimb consists of: Humerus, radius and ulna, carpals, metacarpals, digits or phalanges .The top of the humerus moves against (articulates with) the scapula at the shoulder joint. By changing the number, size and shape of the various bones, fore limbs have evolved to fit different ways of life. They have become wings for flying in birds and bats, flippers for swimming in whales, seals and porpoises, fast and efficient limbs for running in horses and arms and hands for holding and manipulating in primates.

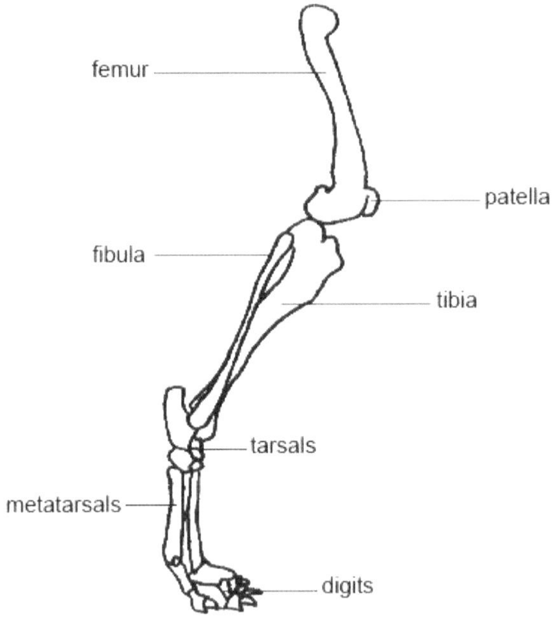

Hindlimb of a dog

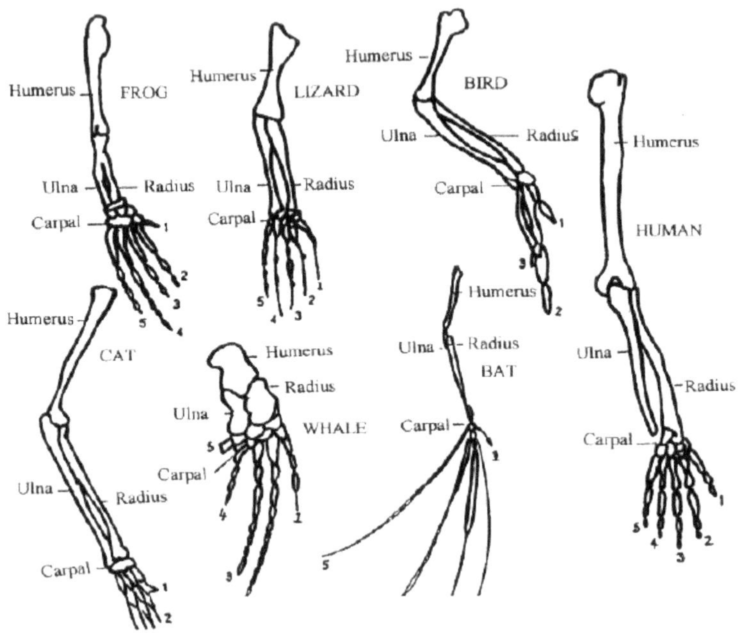

Various vertebrate limbs

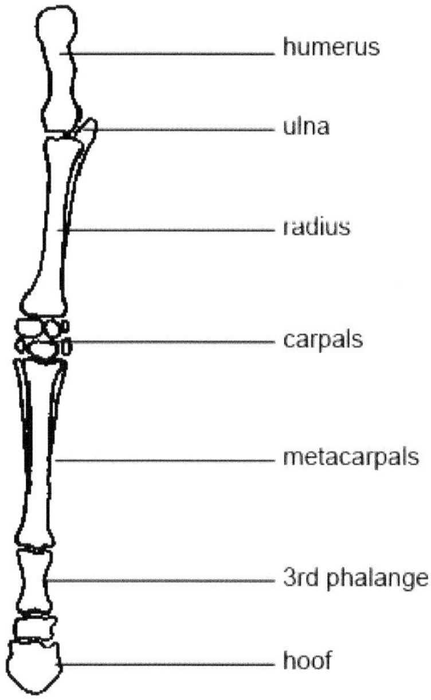

Forelimb of a horse

In the horse and other equines, the third toe is the only toe remaining on the front and real limbs. Each toe is made up of a proximal phalange, a middle phalange, and distal phalange (and some small bones often referred to as sesamoids. In this image, the proximal phalange is labeled P3 and the distal phalange is labeled hoof.

The legs of the horse are highly adapted to give it great galloping speed over long distances. The bones of the lower leg and foot are greatly elongated and the hooves are actually the tips of the third fingers and toes, the other digits having been lost or reduced .

The Hind Limb

The hind limbs have a similar basic pattern to the forelimb. They consist of: femur, tibia and fibula, tarsals, metatarsals, digits or phalanges .The top of the femur moves against (articulates with) the pelvis at the hip joint.

The Girdles

The girdles pass on the "push" produced by the limbs to the body. The shoulder girdle or scapula is a triangle of bone surrounded by the muscles of the back but not connected directly to the spine (see diagram 6.1). This arrangement helps it to cushion the body when landing after a leap and gives the forelimbs the flexibility to manipulate food or strike at prey. Animals that use their forelimbs for grasping, burrowing or climbing have a well-developed clavicle or collar bone. This connects the shoulder girdle to the sternum. Animals like sheep, horses and cows that use their forelimbs only for supporting the body and locomotion have no clavicle. The pelvic girdle or hipbone attaches the sacrum and the hind legs. It transmits the force of the leg-thrust in walking or jumping directly to the spine.

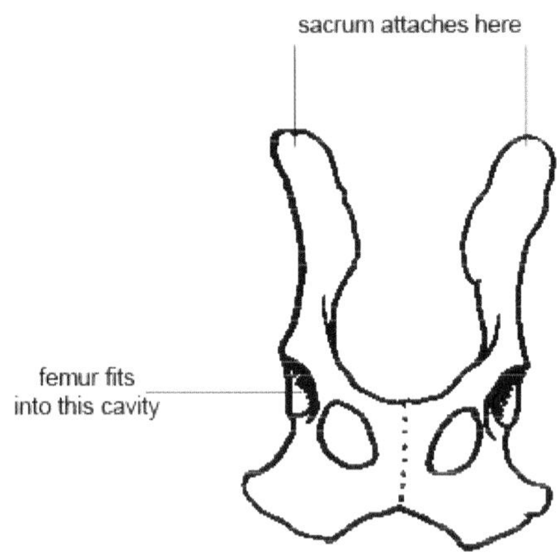

The pelvic girdle

Categories of Bones

People who study skeletons place the different bones of the skeleton into groups according to their shape or the way in which they develop. Thus we have long bones like the femur, radius and finger bones, short bones like the ones of the wrist and ankle, irregular bones like the vertebrae and flat bones like the shoulder blade and bones of the skull. Finally there are bones that develop in tissue separated from the main skeleton. These include sesamoid bones which include bones like the patella or kneecap that develop in tendons and visceral bones that develop in the soft tissue of the penis of the dog and the cow's heart.

Bird Skeletons

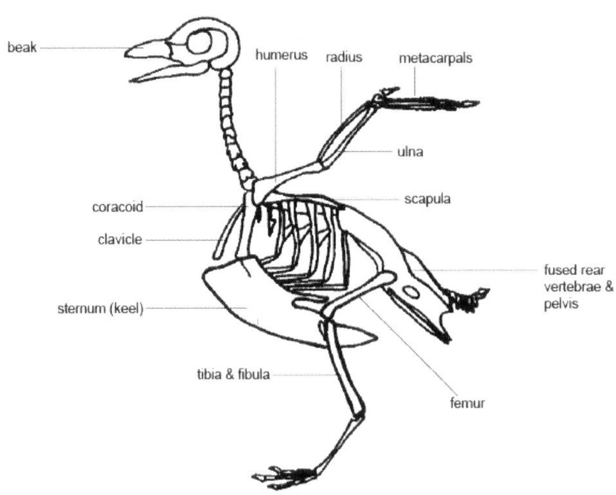

A bird's skeleton

Although the skeleton of birds is made up of the same bones as that of mammals, many are highly adapted for flight. The most noticeable difference is that the bones of the forelimbs are elongated to act as wings. The large flight muscles make up as much as 1/5th of the body weight and are attached to an extension of the sternum called the keel. The vertebrae of the lower back are fused to provide the rigidity needed to produce flying movements. There are also many adaptations to reduce the weight of the skeleton. For instance birds have a beak rather than teeth and many of the bones are hollow.

The Structure of Long Bones

A long bone consists of a central portion or shaft and two ends called epiphyses. Long bones move against or articulate with other bones at joints and their ends have flattened surfaces and rounded protuberances (condyles) to make this possible. If you carefully examine a long bone you may also see raised or rough surfaces. This is where the muscles that move the bones are attached. You will also see holes (a hole is called a foramen) in the bone. Blood vessels and nerves pass into the bone through these. You may also be able to see a fine line at each end of the bone. This is called the growth plate or epiphyseal line and marks the place where increase in length of the bone occurred.

If you cut a long bone lengthways you will see it consists of a hollow cylinder. The outer shell is covered by a tough fibrous sheath to which the tendons are attached. Under this is a layer of hard, dense compact bone. This gives the bone its strength. The central cavity contains fatty yellow marrow, an important energy store for the body, and the ends are made from honeycomb-like bony material called spongy bone. Spongy bone contains red marrow where red blood cells are made.

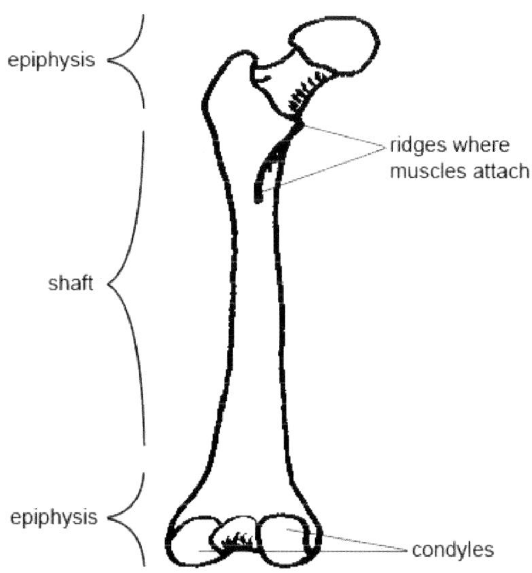

A femur

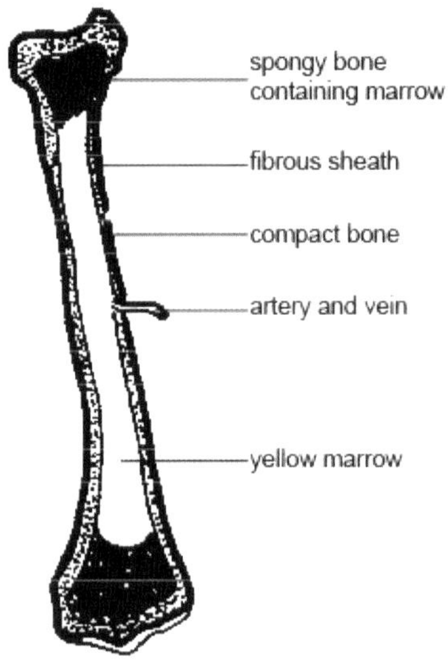

A longitudinal section through a long bone

Compact Bone

Compact bone is not the lifeless material it may appear at first glance. It is a living dynamic tissue with blood vessels, nerves and living cells that continually rebuild and reshape the bone structure as a result of the stresses, bends and breaks it experiences. Compact bone is composed of microscopic hollow cylinders that run parallel to each other along the length of the bone. Each of these cylinders is called a Haversian system. Blood vessels and nerves run along the central canal of each Haversian system. Each system consists of concentric rings of bone material (the matrix) with minute spaces in it that hold the bone cells. The hard matrix contains crystals of calcium phosphate, calcium carbonate and magnesium salts with collagen fibres that make the bone stronger and somewhat flexible. Tiny canals connect the cells with each other and their blood supply.

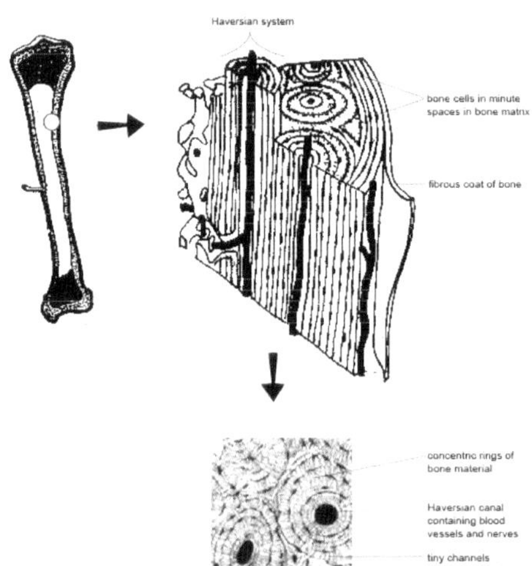

Haversian systems of compact bone

Spongy Bone

Spongy bone gives bones lightness with strength. It consists of an irregular lattice that looks just like an old fashioned loofah sponge (see diagram 6.15). It is found on the ends of long bones and makes up most of the bone tissue of the limb girdles, ribs, sternum, vertebrae and skull. The spaces contain red marrow, which is where red blood cells are made and stored.

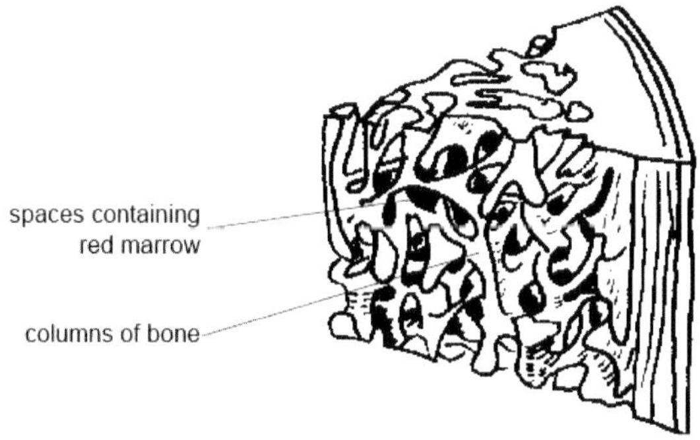

Spongy bone

Bone Growth

The skeleton starts off in the foetus as either cartilage or fibrous connective tissue. Before birth and, sometimes for years after it, the cartilage is gradually replaced by bone. The long bones increase in length at the ends at an area known as the epiphyseal plate where new cartilage is laid down and then gradually converted to bone. When an animal is mature, bone growth ceases and the epiphyseal plate converts into a fine epiphyseal line.

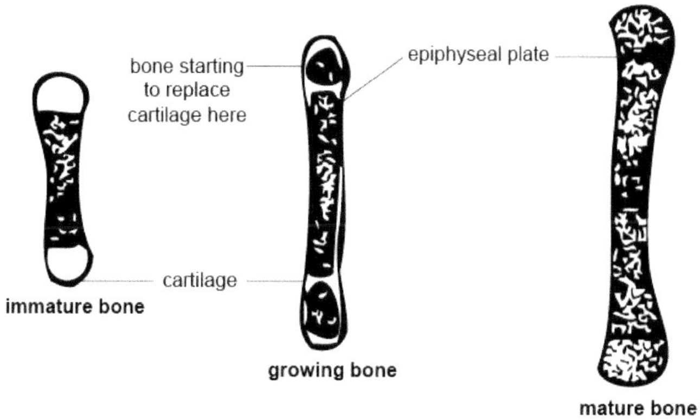

A growing bone

Broken Bones

A fracture or break dramatically demonstrates the dynamic nature of bone. Soon after the break occurs blood pours into the site and cartilage is deposited. This starts to connect the broken ends together. Later spongy bone replaces the cartilage, which is itself replaced by compact bone. Partial healing to the point where some weight can be put on the bone can take place in 6 weeks but complete healing may take 3–4 months.

Joints

Joints are the structures in the skeleton where 2 or more bones meet. There are several different types of joints. Some are immovable once the animal has reached maturity. Examples of these are those between the bones of the skull and the midline joint of the pelvic girdle. Some are slightly moveable like the joints between the vertebrae but most joints allow free movement and have a typical structure with a fluid filled cavity separating the articulating surfaces (surfaces that move against each other) of the two bones. This kind of joint is called a synovial joint (see diagram 6.17). The joint is held together by bundles of white fibrous tissue called ligaments and a fibrous capsule encloses the joint. The inner layers of this capsule secrete the synovial fluid that acts as a lubricant. The articulating surfaces of the bones are covered with cartilage that also reduces friction and some joints, e.g. the knee, have a pad of cartilage between the surfaces that articulate with each other.

The shape of the articulating bones in a joint and the arrangement of ligaments determine the kind of movement made by the joint. Some joints only allow a to and from gliding movement e.g. between the ankle and wrist bones; the joints at the elbow, knee and fingers are hinge joints and allow movement in two dimensions and the axis vertebra pivots on the atlas vertebra. Ball and socket joints, like those at the shoulder and hip, allow the greatest range of movement.

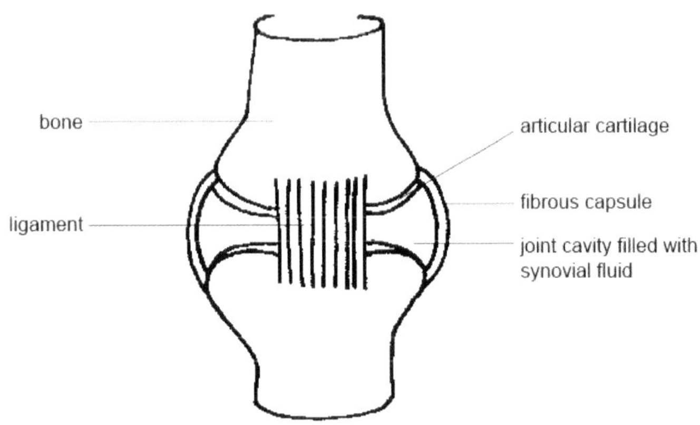

A synovial joint

Common Names of Joints

Some joints in animals are given common names that tend to be confusing. For example:

1. The joint between the femur and the tibia on the hind leg is our knee but the stifle in animals.

2. Our ankle joint (between the tarsals and metatarsals) is the hock in animals

3. Our knuckle joint (between the metacarpals or metatarsals and the phalanges) is the fetlock in the horse.

4. The "knee" on the horse is equivalent to our wrist (ie on the front limb between the radius and metacarpals).

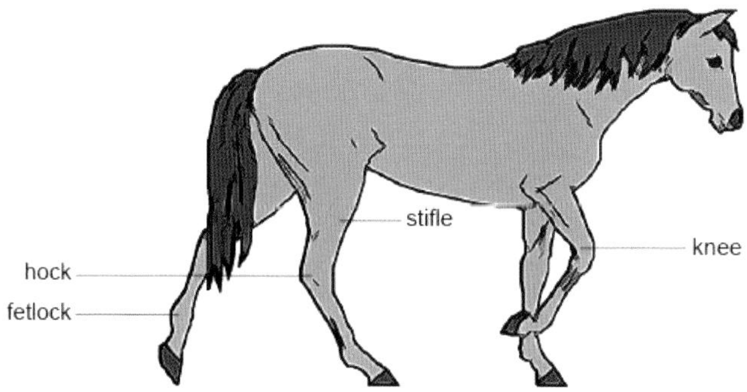

The names of common joints of a horse

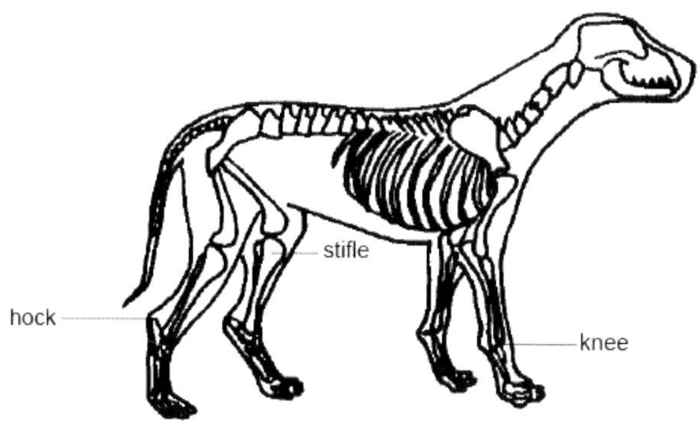

The names of common joints of a dog

Locomotion

Different animals place different parts of the foot or forelimb on the ground when walking or running.

Humans and bears put the whole surface of the foot on the ground when they walk. This is known as plantigrade locomotion. Dogs and cats walk on their toes (digitigrade locomotion) while horses and pigs walk on their "toenails" or hoofs. This is called unguligrade locomotion.

1. Plantigrade locomotion (on the "palms of the hand) as in humans and bears

2. Digitigrade locomotion (on the "fingers") as in cats and dogs

3. Unguligrade locomotion (on the "fingernails") as in horses

Animal Nervous System

The nervous system is one of the leading integrating systems, which carry the interaction and integration of all systems within the whole organism, maintain homeostasis. In addition, the nervous system connects the body with his environment, participating in the process of adaptation. The principle of the nervous system is the doctrine of reflex.

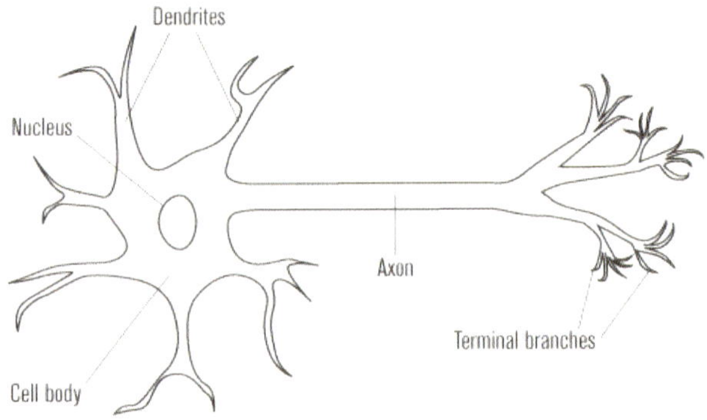

Nervous system of animals

Reflex - is the body's response to irritation, carried out with the participation and control of the central nervous system. The nervous system is the principle of feedback. In the study of the nervous system representation of the reflex arc has changed the concept of the reflex ring, comprising a receptor sensitive neuron, motor neuron, effector and receptor again, but different. Structural and functional unit of the nervous system are nerve cells - neurons with their processes, synapses (the place of transition from one cell to another, where the electrical signal is transformed into chemical and vice versa). Among neurons are three main types: sensory (afferent), false (associative) and motor (efferent). Processes of nerve cells are the basis of the nerves that are also divided into sensory (somatic and visceral) motor (somatic and autonomic) and trophic.

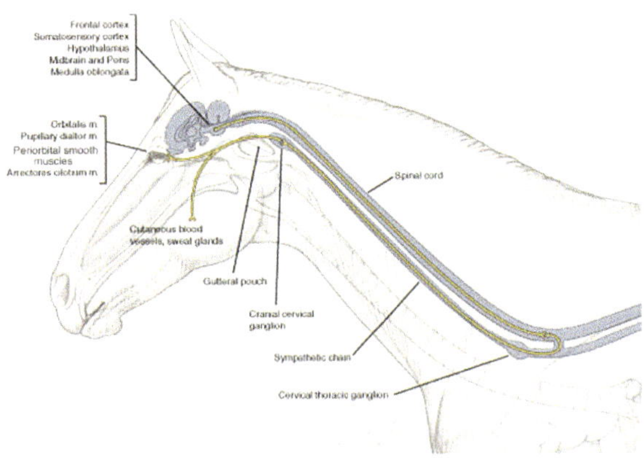

Nervous system of horse

For topographical principle nervous system is divided into the central nervous system (brain and spinal cord) and peripheral nervous system (spinal and cranial nerves with their roots, branches, ganglia and endings). For functional grounds entire nervous system is divided into somatic (innervates musculoskeletal and skin) and autonomic (innervates internal organs and blood vessels, ie smooth muscle tissue).

Central Nervous System

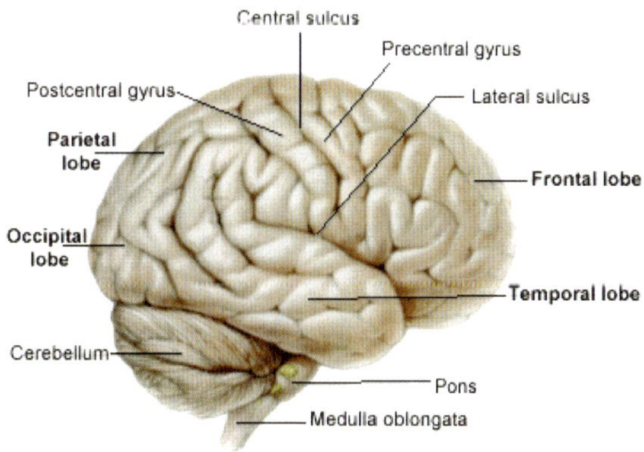

Brain, Nervous system of animals

The central nervous system is a tubular structure, that contains a cavity that is in the spinal cord called the central spinal canal and the brain - the ventricles. Completed this cavity cerebrospinal fluid. Central Division of the nervous system has three meninges: hard, soft and gossamer, between which are formed podobolochechnye space is also filled with cerebrospinal fluid. Central Division of the nervous system is formed by gray and white brain matter.

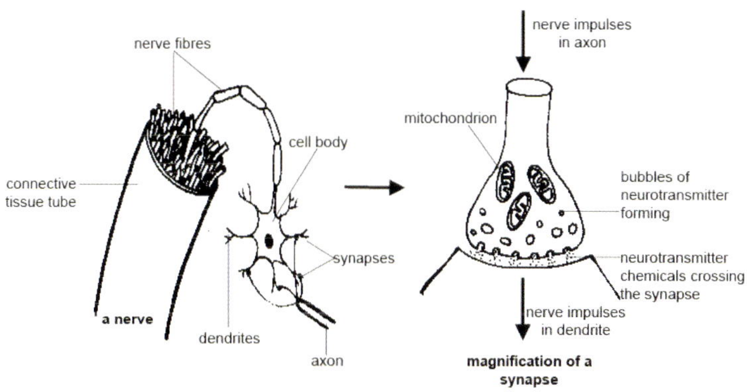

Nerve, Nervous system of animals

Grey medulla represented bodies and processes of nerve cells containing nerve centers and is around cavities (only in cerebral hemisphere and cerebellum gray matter at the periphery and is called the cortex). White medulla established branches of nerve cells and represents the pathways of the brain, located at the periphery of the brain (except cerebellum and cortex). The spinal cord (medulla spinalis) located in the spinal canal and close segments by moving away spinal nerves. The spinal cord is divided into the same sections as the spine: cervical, thoracic, lumbar and sacral. Lumbosacral spinal cord with nerves off of it forms a ponytail. Gray matter of the spinal cord forms the horns of different functions: dorsal horn is sensitive ventral - motor, and thoracolumbar spine is lateral horns containing centers of the sympathetic nervous system. White matter of spinal cord forms: dorsal (sensory pathways), ventral (motor pathways) and lateral (mixed).Spinal cord following functions: conductor (includes pathways that connect different parts of the central nervous system), reflex (contains some centers reflexes), sensitive (associated with the perception of pain). Brain (encephalon) is divided into diamond-shaped and large. Rhomboid brain is divided into the medulla (medulla oblongata) and back, which consists of the cerebellum (cerebellum) and bridge (pons). When the cerebellum is the fourth cerebral ventricle.

In the medulla oblongata are centers of cardiovascular activity, respiration, vomiting center, saliva and tearing, motility of the stomach and intestines. Since the cerebellum is associated equilibrium function, motor coordination and muscle tone. The bridge consists mainly of white medulla and contains powerful conducting apparatus of the cerebral cortex to the spinal cord. Large brain (cerebrum) consists of medium, intermediate and final. Midbrain (mesencephalon) consists of chetveroholmyya and legs of the brain, separated cerebral (Sylvian) water supply. This section of the brain ensures the implementation of motor reflexes under the influence of light and sound stimuli, and includes motor center spinal cord (red nucleus). Diencephalon (diencephalon) of visual hills (talamus), hypothalamus and epithalamus, contains the third cerebral ventricle. This section includes the brain pathways all kinds of sensitivity, performs endocrine function (pituitary and pineal gland), provides regulation of autonomic functions, is responsible for emotions (fear, anger, joy, satisfaction, etc.). Endbrain (telencephalon) consists of olfactory brain , striatum and skirt. Olfactory structures are located on the basal surface of the brain and include the primary olfactory centers. Striatum - basal ganglia - are centers of unconditioned reflexes and cloak - higher integration centers and conditioned reflexes (higher nervous activity). Eventually the brain are also lateral ventricles (first and second).

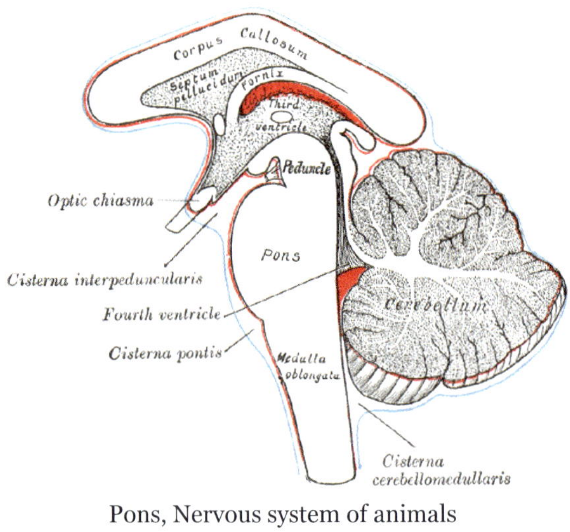

Pons, Nervous system of animals

The Nerve Center

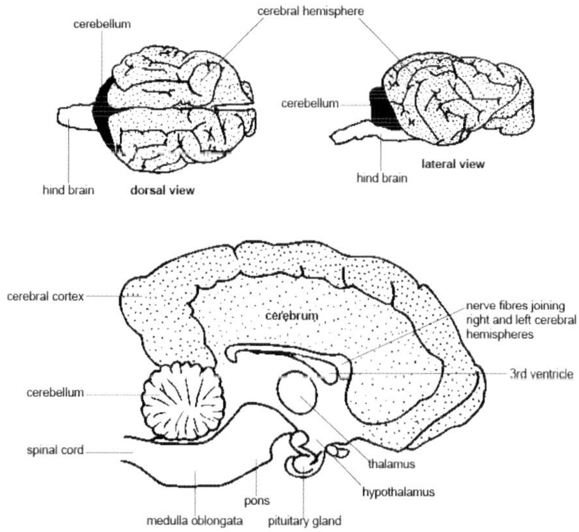

Brain, Nervous system of animals

A group of neurons in the central nervous system involved in the regulation of any function. Neurons that form the nerve center may be located in different parts of the central nervous system.The nerve centers are characterized by a number of properties:

1 unilateral conduction (due to the presence of synapses)

2 slow conduction (due to the large number of synapses in the nerve center)

3 summation of excitation, which can be time and space (defined functional features synapses)

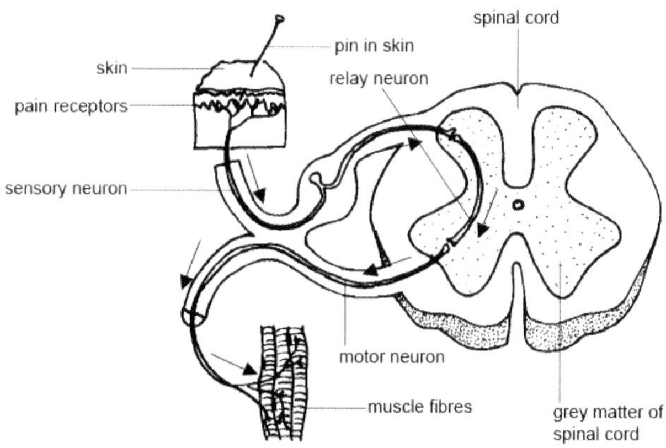

Nervous system of animals

4 irradiation of excitation - excitation of one center causes excitation of another

5 result in nerve centers - excitation delay after the cessation of the stimulus as a result of many nerve connections inside the center

6 facilitate - increase the excitability of the nerve center

7 convergence (convergence - proceeds to motor neuron impulses in several ways)

8 circulation pulses (due to the presence of many nerve connections between neurons in the nerve center)

9 inertia - the ability to maintain long traces of excitement

10 plasticity - the ability to rebuild function, dominant - persistent violation of the center, which occupies a dominant position in the nervous system

11 tone - a state of constant minor offense, fatigue (associated with conducting excitation via synapses due to the high level of metabolism)

12 inhibition - the process weakening or cessation of any activity that may be primary (caused inhibitory neurons) and secondary (occurs in the same neurons in which excitation occurs, under certain conditions).

Higher Nervous Activity

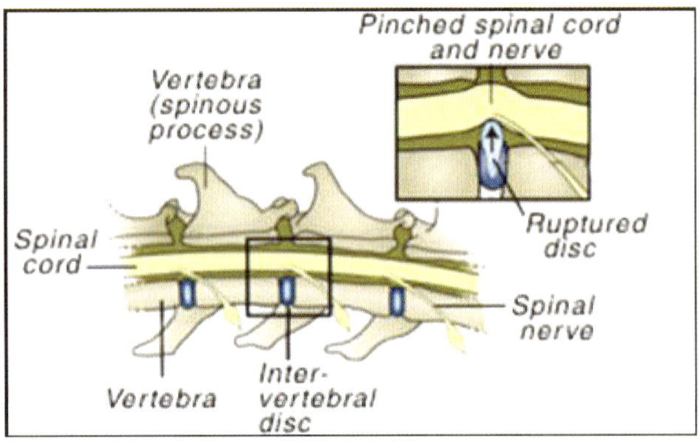

spinal cord, Nervous system of animals

Activity of the higher parts of the nervous system, which in mammals are centers inherent in the cloak of the forebrain. Higher nervous activity is manifested in the ability to develop conditioned reflexes that unlike unconditioned reflexes (the result of subcortical brain) is acquired, transitory, individual, do not have permanent reflex arc produced gradually from unconditioned reflexes handed down from generation to generation in aid learning (imitative reflexes).The biological significance of conditioned reflexes is their participation in the process of behavioral adaptation, so if you change conditions in the cerebral cortex arises process of inhibition of conditioned reflexes. Inhibition divided into conventional (occurs in those centers that himself reflex) and unconditional (provided from outside and may be external and transcendent). For a conditioned reflex to compliance with certain conditions: repeated coincidence in time conditional and unconditional stimuli, conditioned stimulus should begin to act for a few seconds before the unconditioned stimulus.As a result, the two centers in the brain there is a temporary connection as unconditioned reflex center in the brain is dominant, he pulls over excitement that occurs in the center of the conditioned reflex.

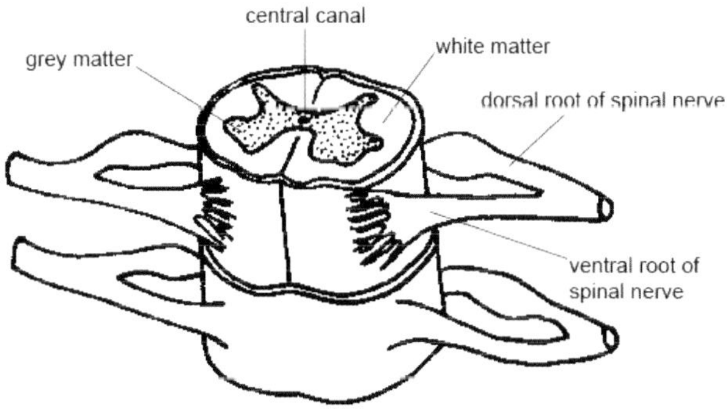

Spinal canal, Nervous system of animals

The ability to develop conditioned reflexes and speed their development formed the basis of the doctrine of types of higher nervous activity, which is characterized by three parameters: the strength of nerve processes (efficiency of brain cells), balance (the ratio between the strength of excitation and inhibition) and mobility (rate of change of excitation and braking).

Established four types of higher nervous activity:

1. Strong, balanced, agile (sanguine).
2. Strong, balanced, inert (phlegmatic).
3. Strong unbalanced (choleric).
4. Weak (melancholic).

The basis of animal behavior are instincts - a system of chain complex unconditioned reflexes that are in the process of life piling weight of conditioned reflexes.The science that studies animal behavior, called ethology.

Peripheral Nervous System

The peripheral nervous system consists of permeate dog body nerves that conduct impulses from receptors to the brain and from the brain to your body. In the course of nerve ganglia are located.Nerves are divided on spinal and cranial. Segments by spinal nerves branch off from the spinal cord with which they are associated spines (dorsal - ventral sensitive and - motor). On the dorsal spine is spinal ganglion. On exit from the spinal canal nerve divided into dorsalnuyu and ventral branches. Ventral branches of nerves, except infants, involved in the formation of nerve plexuses (cervical, shoulder - innervates thoracic limb and lumbosacral - innervates abdominal stack, pelvic limb, external genitalia and udder in females).

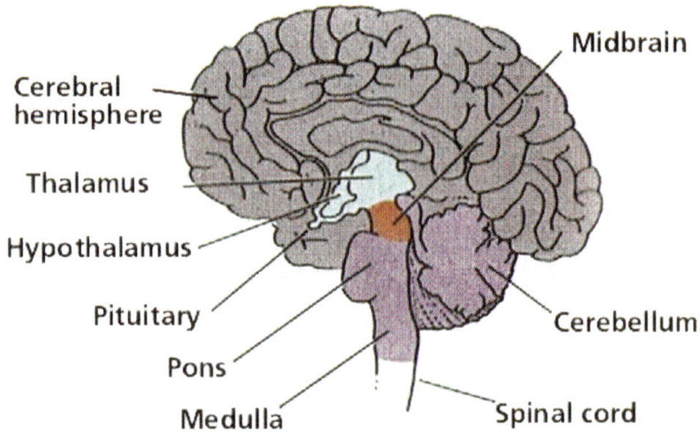

Central Nervous system of animals

Among the brachial plexus nerves largest reach of radial nerve, which innervates all extensor elbow, carpal and finger joints, elbow and median nerves that innervate the carpal and fleksorov finger joints. The median nerve reaches third phalanges.

Among the nerves of the lumbosacral plexus is the largest femoral nerve (n. femoralis), which innervates the quadriceps muscle of the thigh and is located in the femoral canal on the medial thigh and sciatic nerve (n. ischia-dicus), which innervates almost all pelvic limb and is divided into the tibial (n. tibialis) and fibula (n. reroneus) nerves.

The cranial nerves are 12 pairs presented and its functionality are divided into three groups:

1. Sensitive - olfactory (I pair associated with the forebrain), optic (II vapor associated with intermediate brain) and preddverno-cochlear (VIII pair associated with the medulla oblongata).
2. Motor - oculomotor (III couple departs from the midbrain, innervates the muscles of the eyeball), block (IV pair departs from the midbrain, innervates the muscles of the eyeball), abducent (VI pair departs from the medulla innervates m 'muscles of the eyeball), optional (XI pair departs from the medulla and spinal cord, innervates the trapezius and brachiocephalic muscle), hypoglossal (XII couple departs from the medulla innervates muscles of speech).
3. Mixed - trigeminal nerve (V pair departs from cerebral bridge provides sensitive innervation of all regions of the head and motor innervation of masticatory muscles), personnel (VII pair departs from the medulla oblongata, provides innervation of facial muscles, taste buds tongue, salivary and lacrimal glands), glossopharyngeal nerve (IX pair departs from the medulla innervates the region of the pharynx and tongue, parotid salivary gland), vagus (X couple departs from the medulla oblongata, belongs to the parasympathetic nervous system).

Autonomic Nervous System

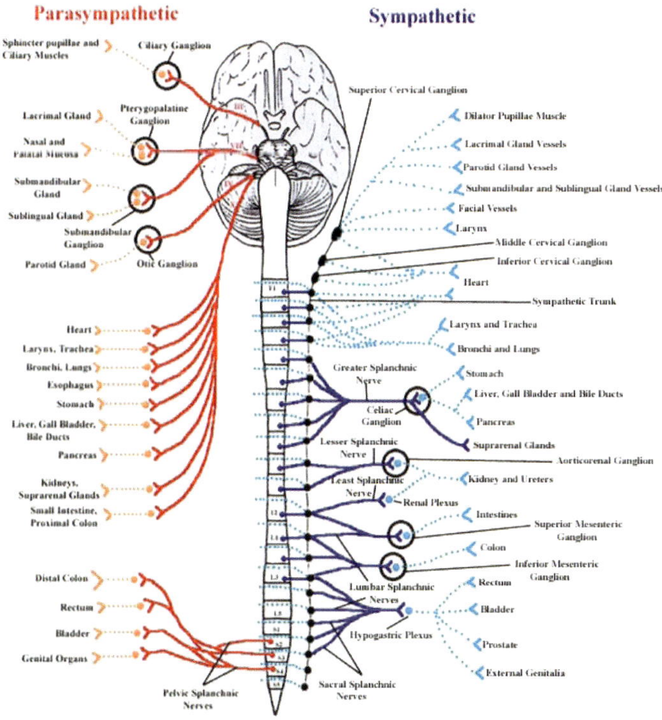

Autonomic nervous system, Nervous system of animals

The autonomic nervous system is divided into two divisions: the sympathetic and parasympathetic, which differ from one location centers and ganglia objects innervation and structure of the motor reflex path.

Sympathetic

Sympathetic division innervates vascular smooth muscle. Its centers are located in the lateral horns of the thoracolumbar spinal cord. Ganglia are on the vertebral bodies (paravertebral ganglia form the border sympathetic trunk) or close to them (prevertebralnye ganglia: cranial cervical, semilunar, caudal mesenteric).

Preganglionic (douzlovye) fibers are short, postganglionic (posleuzlovye) long and part of the cranial nerves, spinal nerves or form a special sympathetic nerves.

Parasympathetic

Parasympathetic division innervates smooth muscles of internal organs and glands. Its centers are located in the middle (hence innervated sphincter pupil), medulla (hence innervated lacrimal and salivary glands, organs and glands neck, chest and abdominal cavities) of the brain and in the lumbosacral spinal cord (hence the innervated organs and glands pelvic cavity). Preganglionic fibers are long and most of them are in the vagus nerve (n. vagus), postganglionic fibers are short. Parasympathetic ganglia located in the wall of the internal organs or surrounding organs (extra-and intramural ganglia).

Senses

A set of anatomical structures that convert the energy of external stimulation in nerve impulse and members of the analyzers. Analyzer - highly complex structures of the body, which distinguish peripheral part - receptor leading portion provided by neural pathways and the central part (area of the brain), where the analysis of nerve impulses. Analyzers different number of properties: high sensitivity, specificity and adaptation, which are caused by the properties of sensory nerve endings - receptors. All receptors are divided into eksteroretseptory (irritable environment and transmit impulses to the cerebral cortex), interoreceptors (excited by stimuli internal environment) and proprioceptors (irritable from musculoskeletal).The structure of the sense organs are eksteroretseptory who perceive the action of temperature, pain and mechanical stimuli (body touch), sound waves (organ of hearing), light (eyesight) and chemical stimuli (olfactory organ and the organ of taste).

Retseptornaya body part represented taste taste buds located in the taste buds of language. Mostly concentrated in kidney valikovidnyh buds (about 200 in each kidney papilla), and horses - in leaf (about 7000 in each kidney). Authority taste best developed in animals with severe chewing surfaces of teeth (the horse, cattle), with age, body taste partially reduced. Different parts of the tongue have affinity to different chemicals.

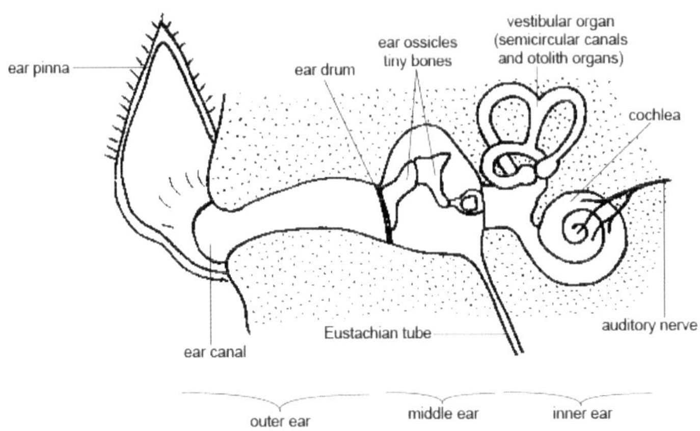

Ear, Nervous system of animals

Olfactory receptor has representation in the olfactory part of the nasal cavity, and pets and in the ventral meatus (Jacobson organ), where the neurosensory epithelium. The degree of development of the olfactory organ pets belong to makrosmatiki. Among pet smell most advanced in predatory and lagomorphs (rabbit).Body contact is represented nerve endings located in the skin and its derivatives. This tactile and pain sensitivity seen the entire surface of the skin, and information on the nature of the items come from endings corolla horn tip toe and myakyshey.Retseptornaya part of the organ of hearing and equilibrium is called ear.

Ear is represented by three parts:

1. The outer ear (catch unit) consists of the auricle with its own muscular and external auditory canal.

2. Middle ear (transmission apparatus) consists of the tympanic cavity (connected to the pharynx by eustachian tube), the eardrum and four auditory ossicles (hammer, anvil, and stapes bones lenticular).

3. Inner ear (takes phone) consists of two mazes - Bone and pereponchatoho, inserted into each other.Bone maze filled perylymfoy and consists of three bone pivkolovyh channel vestibule and cochlea bone.Membranous labyrinth located inside the bone, filled with endolymph and consists of webby pivkolovyh channels, oval and round sacs (located inside the vestibule), membranous cochlea and endolymphatic duct which communicates with podobolochechnye spaces of the brain.

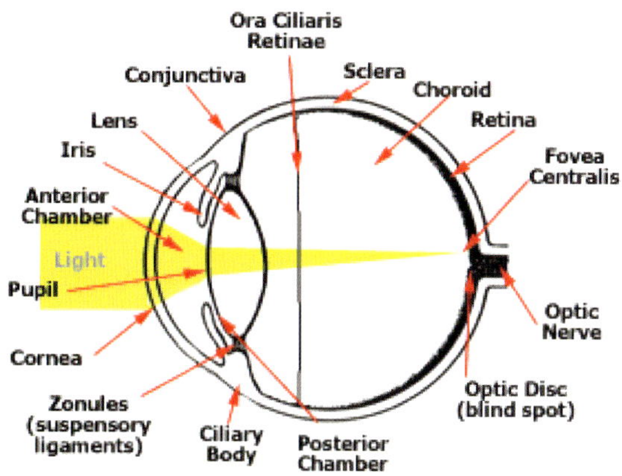

Eyeball, Nervous system of animals

Organ of vision (eye) consists of the eyeball, protective devices and oculomotor apparatus. Eyeball has a three-layered structure and contains svetoprelomlyayuschye environment, which include cornea moist chamber (located in the anterior and posterior chambers of the eye), lens and vitreous.The outer shell of the eyeball (sclera) is protective and its front forming a transparent cornea. The middle layer takes trophic function and forms the iris, in the center of which is a hole - iris and ciliary body (it is thicker resnychnaya muscle, which links to the suspended lens). The inner layer (retina) takes receptor function and contains in its composition layer of rods (responsible for crepuscular vision) and cones (provide a sense of color). Protective devices formed bone orbit, periorbitoy, bands orbital fat, conjunctivitis, lacrimal apparatus, eyelids (top, bottom, and third) and eyelashes. Protective devices and oculomotor apparatus includes seven muscles (four straight and two oblique and ottyahivatel eyeball).

Endocrine System of Animals

The endocrine system is the system of glands, each of which secretes a type of hormone into the bloodstream to regulate the body. While powerful, the endocrine organs are relatively small.

Human endocrine glands are essentially identical to dog and cat endocrine glands, both in structure and function. The endocrine glands include the pituitary gland, thyroid gland, parathyroid glands, pancreas, adrenal glands, ovaries, and testes.

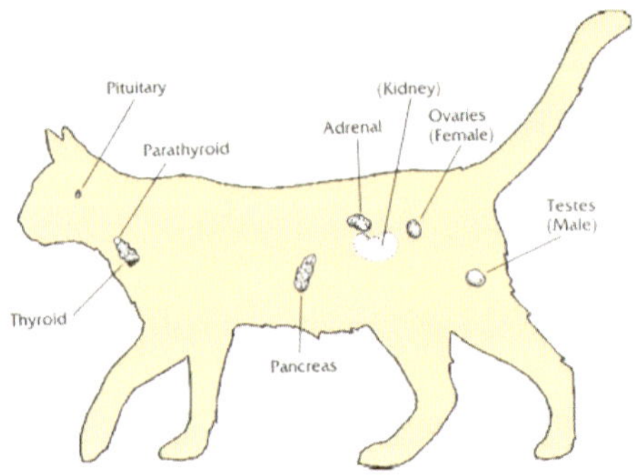

The endocrine system of a cat

Only the larger, more complex animals, such as vertebrates and some arthropods, have special endocrine glands to produce hormones; other animals use nerve cells or tissues such as the gonads. Endocrine glands are another example of a partitioning of functions into separate organs, a system that increases efficiency but that requires a relatively large size to maintain. Greater specialization is also associated with greater difficulties in regenerating lost parts or preventing breakdowns in functions.

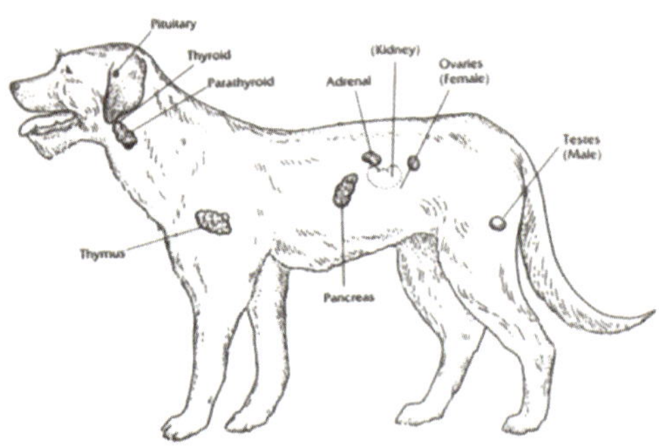

The canine endocrine system

Although the list of hormones found in the mammalian body may seem large, the numbers are surprisingly low for the variety of functions they influence. Which of the multiple functions any one hormone regulates depends on the specificity of the receptors on or within cells. Because all hormones bathe all cells as a result of their transport by the circulatory system, it is more efficient to have a general messenger transported to a cell, where it elicits only one of many possible outcomes. As in the nervous system, the specificity of response lies in the organ that responds and not with the messenger that merely commands action.

Chemicals that allow communication among individuals are called pheromones. Sexual attractants are the most common, but there are many other kinds.

The Pituitary Gland And Hypothalamus

The pituitary gland is a pea-sized structure that is attached by a stalk to the underside of the cerebrum of the brain. It is often called the "master" endocrine gland because it controls many of the other endocrine glands in the body. However, we now know that the pituitary gland is itself controlled by the hypothalamus. This small but vital region of the brain lies just above the pituitary and provides the link between the nervous and endocrine systems. It controls the autonomic nervous system, produces a range of hormones and regulates the secretion of many others from the pituitary gland.

The pituitary gland is divided into two parts with different functions - the anterior and posterior pituitary .

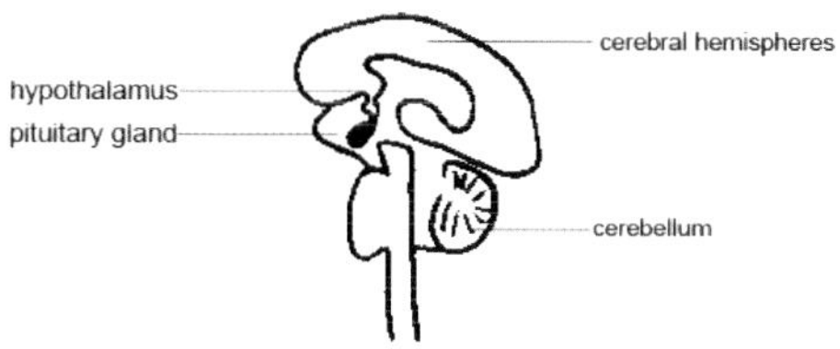

The position of the pituitary gland and hypothalamus

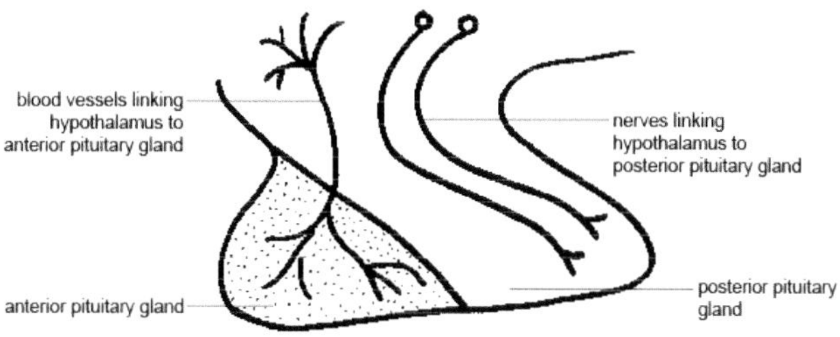

The anterior and posterior pituitary

The anterior pituitary gland secretes hormones that regulate a wide range of activities in the body. These include:

1. Growth hormone that stimulates body growth.
2. Prolactin that initiates milk production.
3. Follicle stimulating hormone (FSH) that stimulates the development of the follicles of the ovaries. These then secrete oestrogen (see chapter 6).
4. melanocyte stimulating hormone (MSH) that causes darkening of skin by producing melanin.

5. lutenizing hormone (LH) that stimulates ovulation and production of progesterone and testosterone.

The posterior pituitary gland:

1. Antidiuretic Hormone (ADH), regulates water loss and increases blood pressure.
2. Oxytocin, milk.

The Pineal Gland

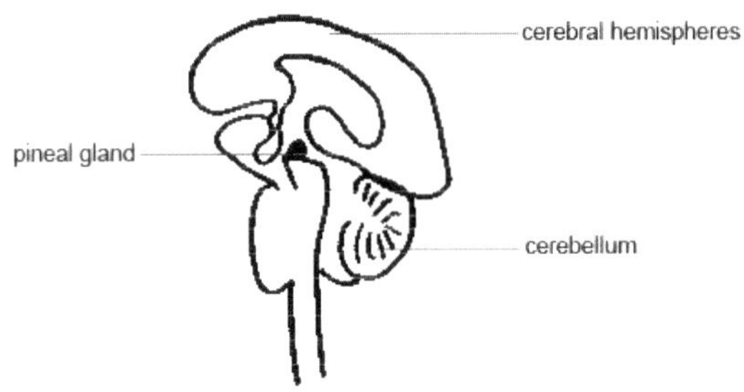

The pineal gland

The pineal gland is found deep within the brain. It is sometimes known as the 'third eye" as it responds to light and day length. It produces the hormone melatonin, which influences the development of sexual maturity and the seasonality of breeding and hibernation. Bright light inhibits melatonin secretion Low level of melatonin in bright light makes one feel good and this increases fertility. High level of melatonin in dim light makes an animal tired and depressed and therefore causes low fertility in animals.

The Thyroid Gland

The thyroid gland is situated in the neck, just in front of the windpipe or trachea. It produces the hormone thyroxine, which influences the rate of growth and development of young animals. In mature animals it increases the rate of chemical reactions in the body.

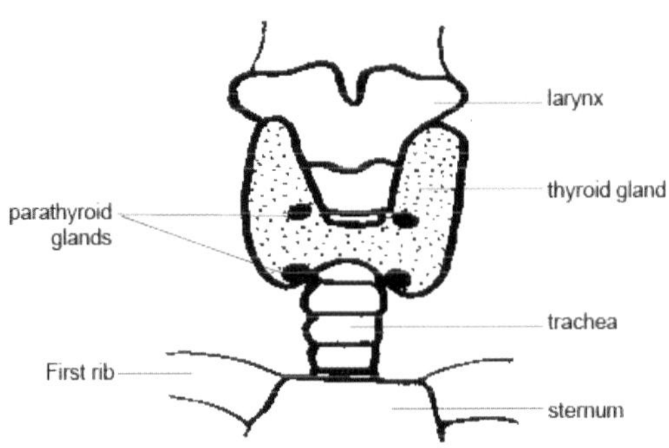

The thyroid and parathyroid glands

Thyroxine consists of 60% iodine and too little in the diet can cause goitre, an enlargement of the thyroid gland. Many inland soils in New Zealand contain almost no iodine so goitre can be common in stock when iodine supplements are not given. To add to the problem, chemicals called goitrogens that occur naturally in plants like kale that belong to the cabbage family, can also cause goitre even when there is adequate iodine available.

The Parathyroid Glands

The parathyroid glands are also found in the neck just behind the thyroid glands .They produce the hormone parathormone that regulates the amount of calcium in the blood and influences the excretion of phosphates in the urine.

The Adrenal Gland

The adrenal glands are situated on the cranial surface of the kidneys. There are two parts to this endocrine gland, an outer cortex and an inner medulla.

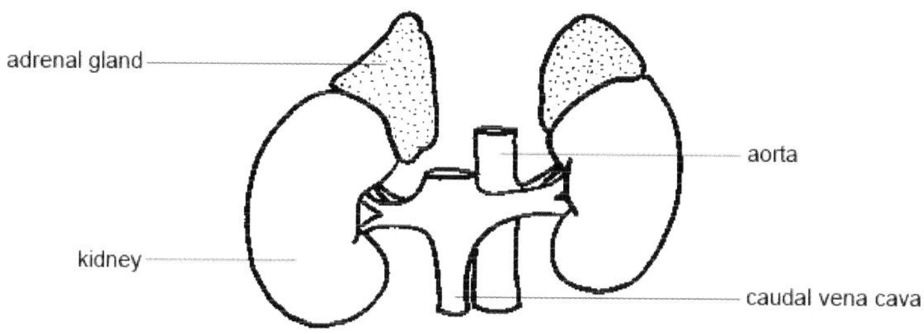

The adrenal glands

The adrenal cortex produces several hormones. These include:

1. Aldosterone that regulates the concentration of sodium and potassium in the blood by controlling the amounts that are secreted or reabsorbed in the kidney tubules.

2. Cortisone and hydrocortisone (cortisol) that have complex effects on glucose, protein and fat metabolism. In general they increase metabolism. They are also often administered to animals to counteract allergies and for treating arthritic and rheumatic conditions. However, prolonged use should be avoided if possible as they can increase weight and reduce the ability to heal.

3. Male and female sex hormones similar to those secreted by the ovaries and testes.

The hormones secreted by the adrenal cortex also play a part in "general adaptation syndrome" which occurs in situations of prolonged stress.

The adrenal medulla secretes adrenalin (also called epinephrine). Adrenalin is responsible for the so-called flight fight, fright response that prepares the animal for emergencies. Faced with a perilous situation the animal needs to either fight or make a rapid escape. To do either requires instant energy, particularly in the skeletal muscles. Adrenaline increases the amount of blood reaching them by causing their blood vessels to dilate and the heart to beat faster. An increased rate of breathing increases the amount of oxygen in the blood and glucose is released from the liver to provide the fuel for energy production. Sweating increases to keep the muscles cool and the pupils of the eye dilate so the animal has a wide field of view. Functions like digestion and urine production that are not critical to immediate survival slow down as blood vessels to these parts constrict.

Note that the effects of adrenalin are similar to those of the sympathetic nervous system.

The Pancreas

In most animals the pancreas is an oblong, pinkish organ that lies in the first bend of the small intestine. In rodents and rabbits, however, it is spread thinly through the mesentery and is sometimes difficult to see.

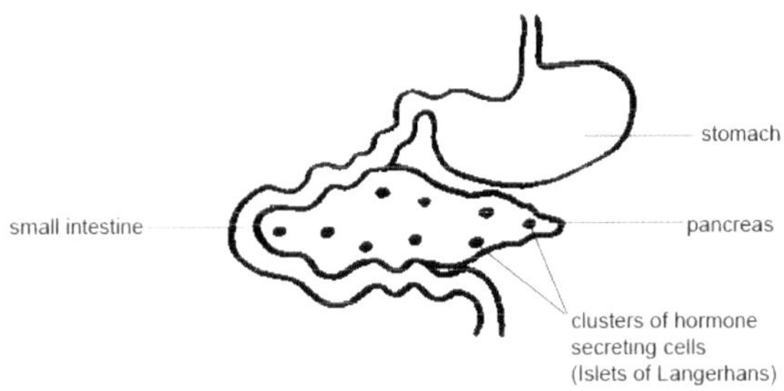

The pancreas

Most of the pancreas acts as an exocrine gland producing digestive enzymes that are secreted into the small intestine. The endocrine part of the organ consists of small clusters of cells (called Islets of Langerhans) that secrete the hormone insulin. This hormone regulates the amount of glucose in the blood by increasing the rate at which glucose is converted to glycogen in the liver and the movement of glucose from the blood into cells.

In diabetes mellitus the pancreas produces insufficient insulin and glucose levels in the blood can increase to a dangerous level. A major symptom of this condition is glucose in the urine.

The Ovaries

The ovaries, located in the lower abdomen, produce two important sex hormones.

1. The follicle cells, under the influence of FSH, produce oestrogen, which stimulates the development of female sexual characteristics - the mammary glands, generally smaller build of female animals etc. It also stimulates the thickening of the lining of the uterus in preparation for pregnancy.

2. Progesterone is produced by the corpus luteum, the endocrine gland that develops in the empty follicle following ovulation. It promotes the further preparation of the uterine lining for pregnancy and prevents the uterus contracting until the baby is born.

Water/Vascular Systems

Animals live in an aquatic environment even on land. Each cell is in contact with the ocean or its aqueous equivalent, which carries food and oxygen to the cells of the animal and carries its metabolic wastes away. The water/vascular systems found in animals vary from the nonexistent to the complex, with the complexity correlated with body size and level of activity. Smaller animals simply use the fluid-filled coelom for transport. Increasing size, however, places too many cells beyond diffusion distance from either the coelom or the outside. A muscular pump attached to muscular vessels has arisen in larger animals to move the interstitial fluid surrounding the cells. Most animals have open circulatory systems. Those few animals with closed circulatory systems have a continuous series of vessels to circulate fluid to the vicinity of all cells, whereas those with open systems have vessels only near the heart. (Actually, no system is entirely closed or open.) In open systems the interstitial fluid and the circulatory fluid are the same, but in closed systems the two fluids can differ considerably in composition.

Closed circulatory systems have several advantages that make them more appropriate than open systems for large, active animals: active animals, in fact, tend to possess closed systems even though their relatives may not. For example, cephalopods, alone among the mollusks, and nemerteans, the most active of acoelomates, have closed systems, as do all annelids and vertebrates. Decapod crustaceans, the largest living arthropods, have nearly closed systems. The most fully open systems have a heart with a few vessels leading from it, while fully closed systems both leak fluid (which is reclaimed by the open lymphatic system) and have open sections. For example, blood flow in the vertebrate liver is partly open.

In closed systems, blood flow can be both higher and directed more often to tissues that require a greater perfusion of blood. If blood is confined within discrete vessels, most of which are muscular, contractions can vary the flow rate according to need by altering the amount of constriction. Thus, the heart beats faster during exercise, when the muscles need more oxygen. Fear changes the distribution of blood flow to ready the muscles for possible imminent action. The more muscular arteries, which carry oxygenated blood to the tissues, can proliferate more finely in active tissues so that more cells are closer to the capillaries, where exchange takes place.

Another advantage of a closed system is the ability to carry a high density of oxygen-bearing cells. Such cells cannot flow smoothly through the sometimes tight interstitial spaces and thus are not much used by animals with open systems. A great deal more oxygen, however, can be carried if the oxygen carrier (such as hemoglobin) is packed into cells. The viscosity of the blood is a function of how many discrete particles are contained within it, and size is of little influence. If all the hemoglobin in the blood of humans were released by dissolving the cell membranes, it would be a thick gel unable to flow. Animals with open systems do aggregate their oxygen carriers into giant polypeptides, but single molecules have limits to their size. Myriapods and insects, highly active arthropods with open systems, circumvented this problem by evolving a tracheal system of respiration, as have some other groups: molecular oxygen is carried by branching tubes to within a few cell lengths of any cell.

A few types of cells protect organisms from a potentially hostile outside environment. Internal cells thus can eliminate any unnecessary ancestral life-support components as they specialize for various functions. This cooperation maintains an ideal internal environment for the members of the society of cells but only at the cost of active labour and expenditure of energy. In particular, the proper water/salt balance of the interstitial fluid is crucial to prevent the cell from shrinking or bloating.

Problems of water/salt balance are usually handled by the same system that eliminates the poisonous ammonia derived from metabolizing nitrogen-containing compounds, such as nucleotides or amino acids. Ammonia dissolves readily in water and thus is removed from an animal that needs to rid itself of excess water anyway. (In small animals the ammonia diffuses into the surrounding water.) With large size or a need for water conservation, animals excrete urea, a less toxic compound but one that also contains carbon and oxygen and thus potential energy. Urea also is highly soluble in water, but its low toxicity means that it can be concentrated before being excreted. Terrestrial animals with problems of water conservation either convert urea into uric acid, a solid compound that can be stored indefinitely in the body or voided with the feces, or develop efficient excretory organs (*e.g.,* the mammalian kidney) that can concentrate the urea. Although water balance is usually handled by the kidney, salt balance is sometimes a specialized function of other organs. For example, because freshwater fish tend to lose a great deal of salt through their gills, they simply expend energy to concentrate salt against a gradient at this location.

Animal Reproduction

Reproduction is the process by which cells and organisms produce other cells and organisms of the same kind. Cell reproduction usually involves division of a cell into two identical parts by means of mitosis or into four different parts by meiosis.

- The reproduction of organisms by the union of male and female reproductive cells (gametes) is called sexual reproduction. Most multicellular animals reproduce sexually.

- Reproduction in which offspring are produced by a single parent, without the union of reproductive cells, is called asexual reproduction. The fission (splitting) of bacterial cells is a form of asexual reproduction.)

Asexual Reproduction

Asexual reproduction produces an identical genetic copy of the parent animal instead of merging the genes of two parents. This means that the offspring created by asexual reproduction can be thought of as clones. This type of reproduction can help a population to grow quickly, but it can also leave the species at risk because there is no variation from one animal to the next. This lack of variation can be a threat because an entire population can be wiped out by one thing, like a disease.

Common kinds of asexual reproduction include budding, fragmentation and regeneration. Budding is when a bud grows inside or outside of a mature animal, it is common in some kinds of coral and sponges. Fragmentation is found in some flatworms called planarians, it occurs when a part breaks off or fragments and then grows into another animal.

Regeneration occurs when part of an animal breaks off and the animal grows it back. Sometimes if the part that is broken off is large enough it may also grow into another animal. Echinoderms like starfish are an example of animals that regenerate.

Sexual Reproduction

Sexual reproduction occurs when female reproductive cells are merged with male reproductive cells to form a new animal that is a genetic combination of the parent animals. Most animals reproduce sexually as it allows for greater variation within the species and increases the ability of a species to adapt to changing environments.

Fish and Amphibians

Most fish and amphibians reproduce in an external process. This takes place when the female of the species lays eggs either in open water or a specially prepared nest. The male then fertilises the eggs and they are allowed to grow. The fertilised eggs then hatch animals that finish developing in their watery environments.

Reptiles

Most reptiles also lay eggs but their eggs are fertilised internally. This means that male sperm cells are delivered into the body of the female where they combine with female sex cells to produce an egg or eggs. Many reptiles build nests for their eggs and when the eggs hatch some reptiles will leave their young to fend for themselves while others will help their young to find food and grow.

Birds

Like reptiles, birds lay eggs that have been fertilised internally. One of the reasons that birds lay eggs is that it would be very difficult to fly with a baby growing inside them. Bird eggs need to be kept warm so that the babies develop properly. This is usually done by building a nest and then sitting on top of the eggs.

Once the eggs have hatched, the parents will normally continue to look after the baby birds, feeding them and protecting them until they are big enough to leave the nest.

Mammals

All mammals feed their young with milk from the mother. Most mammals give birth to live young. Monotremes, like the platypus and echidna, are the only mammals to lay eggs. When their young are hatched, they feed on milk from the mother until they have grown enough to find their own food.

Marsupials like, the kangaroo and koala, give birth to partially-developed young called neonates (nee-oh-nayts). These neonates continue to develop and grow in the mother's pouch where they are fed milk. When they are old enough they will leave the mother's pouch and find their own food.

Placental mammals develop inside the body of their mother, in the uterus. They are fed on nutrients from the mother's body through the placenta. When they are fully developed they will be born and then will be fed with milk from their mother. As they grow older they will be less dependent on their mother and will be able to find their own food. Humans and whales are examples of placental mammals.

Animal Reproductive System

Animal reproductive system is any of the organ systems by which animals reproduce.

Unlike most other organ systems, the reproductive systems of higher animals have not generally become more complex than those of lower forms. Asexual reproduction (*i.e.*, reproduction not involving the union of gametes), however, occurs only in the invertebrates, in which it is common, occurring in animals as highly evolved as the sea squirts, which are closely related to the vertebrates. Temporary gonads are common among lower animals; in higher animals, however, gonads are permanent organs. Hermaphroditism, in which one individual contains functional reproductive organs of both sexes, is common among lower invertebrates; yet separate sexes occur in such primitive animals as sponges, and hermaphroditism occurs in animals more highly evolved—*e.g.*, the lower fishes. Gonads located on or near the animal surface are common in the lowest invertebrates, but in higher animals they tend to be more deeply situated and often involve intricate duct systems. In echinoderms, which are among the highest invertebrates, the gonads hang directly into the sea and spill their gametes into the water. In protochordates, gametes are released into a stream of respiratory water that passes directly into the sea. Duct systems of the invertebrate flatworms (Platyhelminthes) are relatively complex, and those of specialized arthropods (*e.g.*, insects, spiders, crabs) are more complex than those of any vertebrate. Copulatory organs occur in flatworms, but copulatory organs are not ubiquitous among vertebrates other than reptiles and mammals. The trend toward fewer eggs and increased parental care in higher animals may account for the relative lack of complexity in the reproductive systems of some advanced forms. Whereas trends toward increasing structural complexity have often been reversed during evolution, reproductive behaviour patterns in many phylogenetic (*i.e.*, evolutionary) lines have become more complicated in order to enhance the opportunity for fertilization of eggs and maximum survival of offspring.

Sexual Reproduction in Mammals

The reproductive organs of mammals produce the gametes (sperm and egg cells), help them fertilize and then support the developing embryo.

The Male Reproductive System

The male reproductive system consists of a pair of testes that produce sperm (or spermatozoa), ducts that transport the sperm to the penis and glands that add secretions to the sperm to make semen.

The various parts of the male reproductive system with a summary of their functions are shown in.

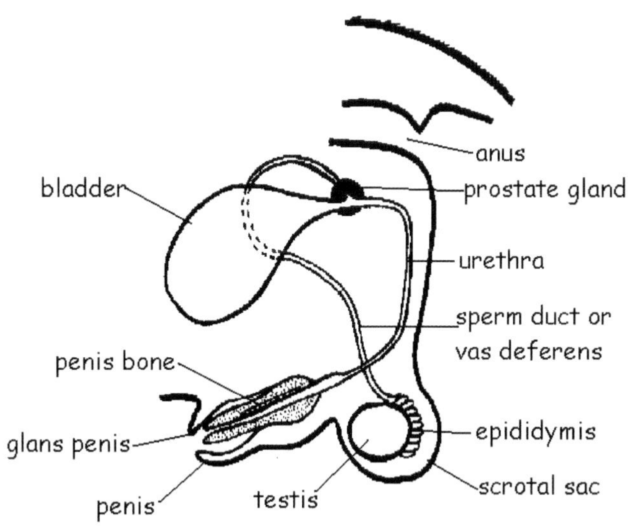

The reproductive organs of a male dog

The Testes

Sperm need temperatures between 2 to 10 degrees Centigrade lower and then the body temperature to develop. This is the reason why the testes are located in a bag of skin called the scrotal sacs (or scrotum) that hangs below the body and where the evaporation of secretions from special glands can further reduce the temperature. In many animals (including humans) the testes descend into the scrotal sacs at birth but in some animals they do not descend until sexual maturity and in others they only descend temporarily during the breeding season. A mature animal in which one or both testes have not descended is called a cryptorchid and is usually infertile.

The problem of keeping sperm at a low enough temperature is even greater in birds that have a higher body temperature than mammals. For this reason bird's sperm are usually produced at night when the body temperature is lower and the sperm themselves are more resistant to heat.

The testes consist of a mass of coiled tubes (the seminiferous or sperm producing tubules) in which the sperm are formed by meiosis. Cells lying between the seminiferous tubules produce the male sex hormone testosterone.

When the sperm are mature they accumulate in the collecting ducts and then pass to the epididymisbefore moving to the sperm duct or vas deferens. The two sperm ducts join the urethra just below the bladder, which passes through the penis and transports both sperm and urine.

Ejaculation discharges the semen from the erect penis. It is brought about by the contraction of the epididymis, vas deferens, prostate gland and urethra.

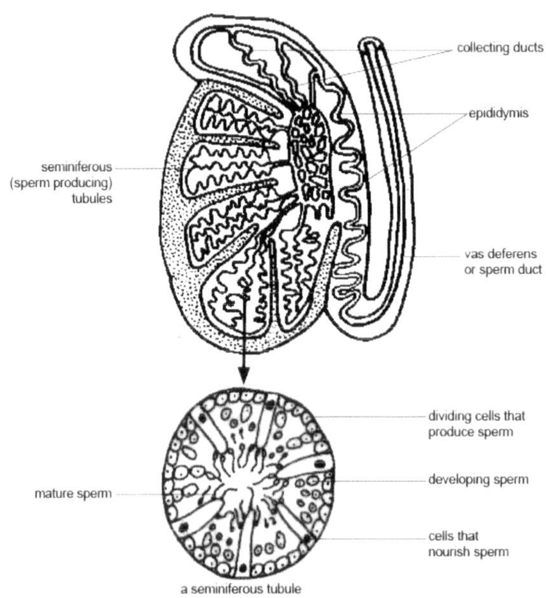

The testis and a magnified seminiferous tubule

Semen

Semen consists of 10% sperm and 90% fluid and as sperm pass down the ducts from testis to penis, (accessory) glands add various secretion...

Accessory Glands

Three different glands may be involved in producing the secretions in which sperm are suspended, although the number and type of glands varies from species to species.

Seminal vesicles are important in rats, bulls, boars and stallions but are absent in cats and dogs. When present they produce secretions that make up much of the volume of the semen, and transport and provide nutrients for the sperm.

The prostate gland is important in dogs and humans. It produces an alkaline secretion that neutralizes the acidity of the male urethra and female vagina.

Cowper's glands (bulbourethral glands) have various functions in different species. The secretions may lubricate, flush out urine or form a gelatinous plug that traps the semen in the female reproductive system after copulation and prevents other males of the same species fertilizing an already mated female. Cowper's glands are absent in bears and aquatic mammals.

The Penis

The penis consists of connective tissue with numerous small blood spaces in it. These fill with blood during sexual excitement causing erection.

Penis form and Shape

Dogs, bears, seals, bats and rodents have a special bone in the penis which helps maintain the erection. In some animals (e.g. the bull, ram and boar) the penis has an "S" shaped bend that allows it to fold up when not in use. In many animals the shape of the penis is adapted to match that of the vagina. For example, the boar has a corkscrew shaped penis, there is a pronounced twist in bulls' and it is forked in marsupials like the opossum. Some have spines, warts or hooks on them to help keep them in the vagina and copulation may be extended to help retain the semen in the female system. Mating can last up to three hours in minks, and dogs may "knot" or "tie" during mating and can not separate until the erection has subsided.

Sperm

Sperm are made up of three parts: a head consisting mainly of the nucleus, a midpiece containing many mitochondria to provide the energy and a tail that provides propulsion.

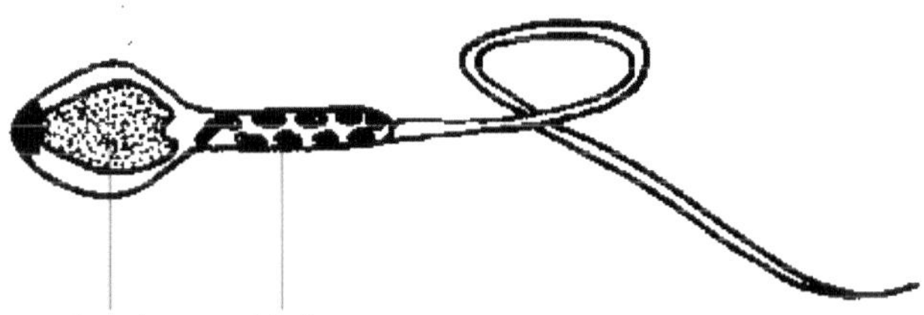

A sperm

A single ejaculation may contain 2-3 hundred million sperm but even in normal semen as many as 10% of these sperm may be abnormal and infertile. Some may be dead while others are inactive or deformed with double, giant or small heads or tails that are coiled or absent altogether.

When there are too many abnormal sperm or when the sperm concentration is low, the semen may not be able to fertilize an egg and the animal is infertile. Make sure you don't confuse infertility with impotence, which is the inability to copulate successfully.

Sperm do not live forever. They have a definite life span that varies from species to species. They survive for between 20 days (guinea pig) to 60 days (bull) in the epididymis but once ejaculated into the female tract they only live from 12 to 48 hours. When semen is used for artificial insemination, storage under the right conditions can extend the life span of some species.

Artificial Insemination

In many species the male can be artificially stimulated to ejaculate and the semen collected. It can then be diluted, stored and used to inseminate females. For example bull semen can be diluted and stored for up to 3 weeks at room temperature. If mixed with an antifreeze solution and stored in "straws" in liquid nitrogen at minus 79°C it will keep for much longer. Unfortunately the semen of chickens, stallions and boars can only be stored for up to 2 days.

Dilution of the semen means that one male can be used to fertilise many more females than would occur under natural conditions. There are also advantages in the male and female not having to make physical contact. It means that owners of females do not have to buy expensive males and the possibility of transmitting sexually transmitted diseases is reduced. Routine examination of the semen for sperm concentration, quality and activity allows only the highest quality semen to be used so a high success rate is ensured.

Since the lifespan of sperm in the female tract is so short and ova only survive from 8 to 10 hours the timing of the artificial insemination is critical. Successful conception depends upon detecting the time that the animal is "on heat" and when ovulation occurs.

The Female Reproductive Organs

The female reproductive system consists of a pair of ovaries that produce egg cells or ova and fallopian tubes where fertilisation occurs and which carry the fertilised ovum to the uterus. Growth of the foetus takes place here. The cervix separates the uterus from the vagina or birth canal, where the sperm are deposited.

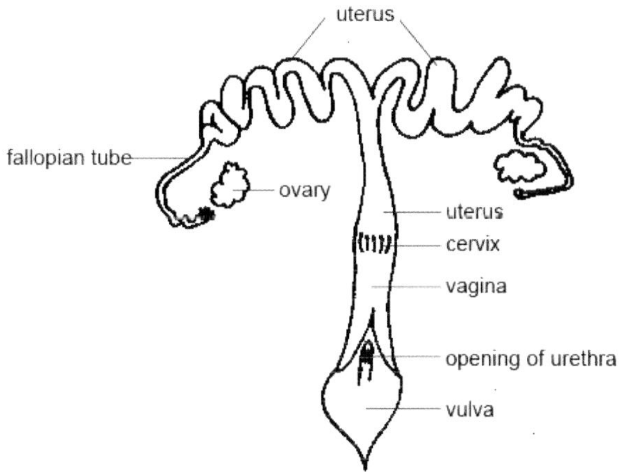

The reproductive system of a female rabbit

Note that primates like humans have a uterus with a single compartment but in most mammals the uterus is divided into two separate parts or horns as shown in diagram.

The Ovaries

Ovaries are small oval organs situated in the abdominal cavity just ventral to the kidneys. Most animals have a pair of ovaries but in birds only the left one is functional to reduce weight.

The ovary consists of an inner region (medulla) and an outer region (cortex) containing egg cells or ova. These are formed in large numbers around the time of birth and start to develop after the animal becomes sexually mature. A cluster of cells called the follicle surrounds and nourishes each ovum.

The Ovarian Cycle

The ovarian cycle refers to the series of changes in the ovary during which the follicle matures, the ovum is shed and the corpus luteum develops.

Numerous undeveloped ovarian follicles are present at birth but they start to mature after sexual maturity. In animals that normally have only one baby at a time only one ovum will mature at once but in litter animals several will. The mature follicle consists of outer cells that provide nourishment. Inside this is a fluid-filled space that contains the ovum.

A mature follicle can be quite large, ranging from a few millimetres in small mammals to the size of a golf ball in large animals. It bulges out from the surface of the ovary before eventually rupturing to release the ovum into the abdominal cavity. Once the ovum has been shed, a blood clot forms in the empty follicle. This develops into a tissue called the corpus luteum that produces the hormone progesterone. If the animal becomes pregnant the corpus luteum persists, but if there is no pregnancy it degenerates and a new ovarian cycle usually.

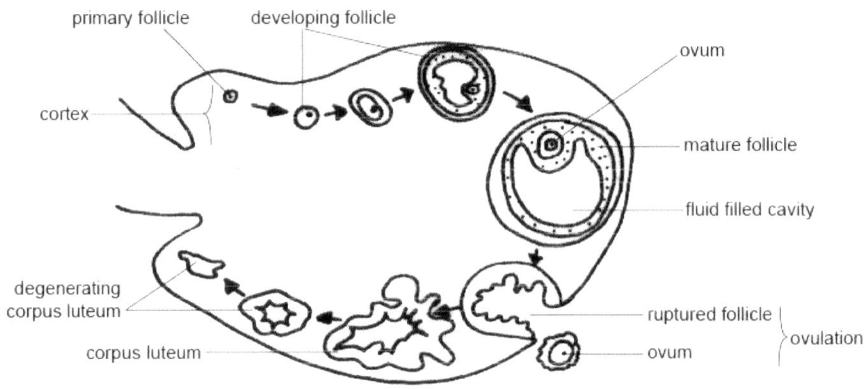

The ovarian cycle showing from the top left clockwise: the maturation of the ovum over time, followed by ovulation and the development of the corpus luteum in the empty follicle

The Ovum

When the ovum is shed the nucleus is in the final stages of meiosis (cell division). It is surrounded by few layers of follicle cells and a tough membrane called the zona pellucida.

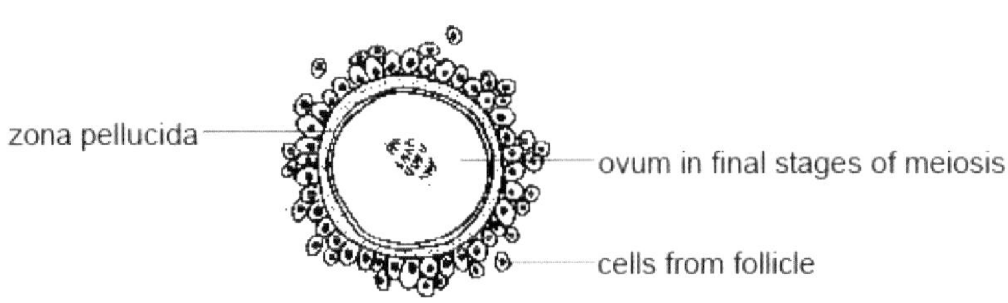

An ovum

The Oestrous Cycle

The oestrous cycle is the sequence of hormonal changes that occurs through the ovarian cycle. These changes influence the behaviour and body changes of the female.

The first hormone involved in the oestrous cycle is follicle stimulating hormone (F.S.H.), secreted by the anterior pituitary gland. It stimulates the follicle to develop.

As the follicle matures the outer cells begin to secrete the hormone oestrogen and this stimulates the mammary glands to develop. It also prepares the lining of the uterus to receive a fertilised egg. Ovulation is initiated by a surge of another hormone from the anterior pituitary, luteinising hormone (L.H.). This hormone also influences the development of the corpus luteum, which produces progesterone, a hormone that prepares the lining of the uterus for the fertilised ovum and readies the mammary glands for milk production. If no pregnancy takes place the corpus luteum shrinks and the production of progesterone decreases. This causes FSH to be produced again and a new oestrous cycle begins.

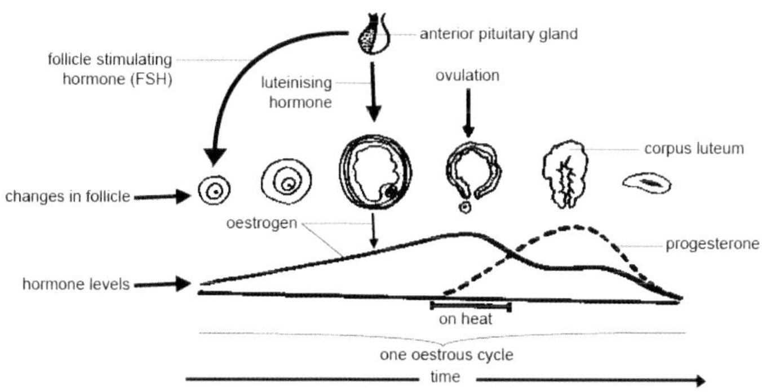

The oestrous cycle

For fertilisation of the ovum by the sperm to occur, the female must be receptive to the male at around the time of ovulation. This is when the hormones turn on the signs of "heat", and she is "in season" or "in oestrous". These signs are turned off again at the end of the oestrous cycle.

During the oestrous cycle the lining of the uterus (endometrium) thickens ready for the fertilised ovum to be implanted. If no pregnancy occurs this thickened tissue is absorbed and the next cycle starts. In humans and other higher primates, however, the endometrium is shed as a flow of blood and instead of an oestrous cycle there is a menstrual cycle.

The length of the oestrous cycle varies from species to species. In rats the cycle only lasts 4–5 days and they are sexually receptive for about 14 hours. Dogs have a cycle that lasts 60–70 days and heat lasts 7–9 days and horses have a 21-day cycle and heat lasts an average of 6 days.

Ovulation is spontaneous in most animals but in some, e.g. the cat, and the rabbit, ovulation is stimulated by mating. This is called induced ovulation.

Signs of Oestrous or Heat

- When on heat a bitch has a blood stained discharge from the vulva that changes a little later to a straw coloured one that attracts all the dogs in the neighbourhood.

- Female cats "call" at night, roll and tread the carpet and are generally restless but will "stand" firm when pressure is placed on the pelvic region (this is the lordosis response).
- A female rat shows the lordosis response when on heat. It will "mount" other females and be more active than normal.
- A cow mounts other cows (bulling), bellows, is restless and has a discharge from the vulva.

Breeding Seasons and Breeding Cycles

Only a few animals breed throughout the year. This includes the higher primates (humans, gorillas and chimpanzees etc.), pigs, mice and rabbits. These are known as continuous breeders.

Most other animals restrict reproduction to one or two seasons in the year-seasonal breeders. There are several reasons for this. It means the young can be born at the time (usually spring) when feed is most abundant and temperatures are favourable. It is also sensible to restrict the breeding season because courtship, mating, gestation and the rearing of young can exhaust the energy resources of an animal as well as make them more vulnerable to predators.

The timing of the breeding cycle is often determined by day length. For example the shortening day length in autumn will bring sheep and cows into season so the foetus can gestate through the winter and be born in spring. In cats the increasing day length after the winter solstice (shortest day) stimulates breeding. The number of times an animal comes into season during the year varies, as does the number of oestrous cycles during each season. For example a dog usually has 2-3 seasons per year, each usually consisting of just one oestrous cycle. In contrast ewes usually restrict breeding to one season and can continue to cycle as many as 20 times if they fail to become pregnant.

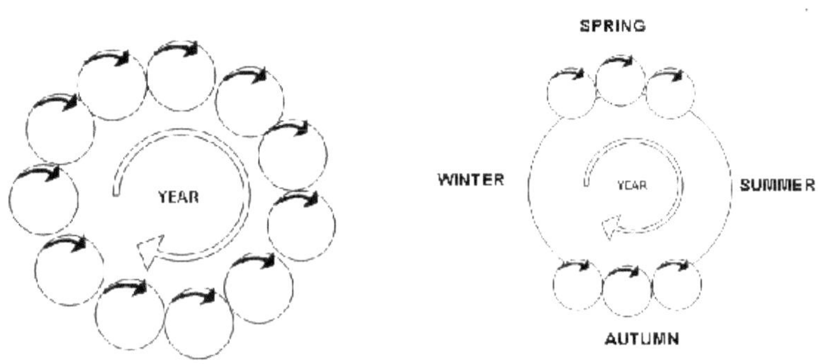

continuous breeders (each circle represents one oestrous cycle)

seasonal breeders

Breeding cycles

Fertilisation and Implantation

Fertilisation

The opening of the fallopian tube lies close to the ovary and after ovulation the ovum is swept into its funnel-like opening and is moved along it by the action of cilia and wave-like contractions of the wall.

Copulation deposits several hundred million sperm in the vagina. They swim through the cervix and uterus to the fallopian tubes moved along by whip-like movements of their tails and contractions of the uterus. During this journey the sperm undergo their final phase of maturation so they are ready to fertilise the ovum by the time they reach it in the upper fallopian tube.

High mortality means only a small proportion of those deposited actually reach the ovum. The sperm attach to the outer zona pellucida and enzymes secreted from a gland in the head of the sperm dissolve this membrane so it can enter. Once one sperm has entered, changes in the zona pellucida prevent further sperm from penetrating. The sperm loses its tail and the two nuclei fuse to form a zygote with the full set of paired chromosomes restored.

Development of the Morula and Blastocyst

As the fertilised egg travels down the fallopian tube it starts to divide by mitosis. First two cells are formed and then four, eight, sixteen, etc. until there is a solid ball of cells. This is called a morula. As division continues a hollow ball of cells develops. This is a blastocyst.

Implantation

Implantation involves the blastocyst attaching to, and in some species, completely sinking into the wall of the uterus.

Pregnancy

The Placenta and Foetal Membranes

As the embryo increases in size, the placenta, umbilical cord and foetal membranes (often known collectively as the placenta) develop to provide it with nutrients and remove waste products. In later stages of development the embryo becomes known as a foetus.

The placenta is the organ that attaches the foetus to the wall of the uterus. In it the blood of the foetus and mother flow close to each other but never mix. The closeness of the maternal and foetal blood systems allows diffusion between them. Oxygen and nutrients diffuse from the mother's blood into that of the foetus and carbon dioxide and excretory products diffuse in the other direction. Most maternal hormones (except adrenaline), antibodies, almost all drugs (including alcohol), lead and DDT also pass across the placenta. However, it protects the foetus from infection with bacteria and most viruses.

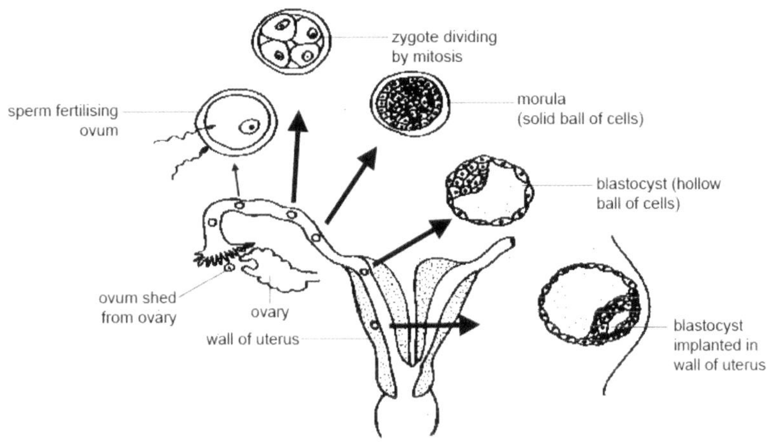

Development and implantation of the embryo

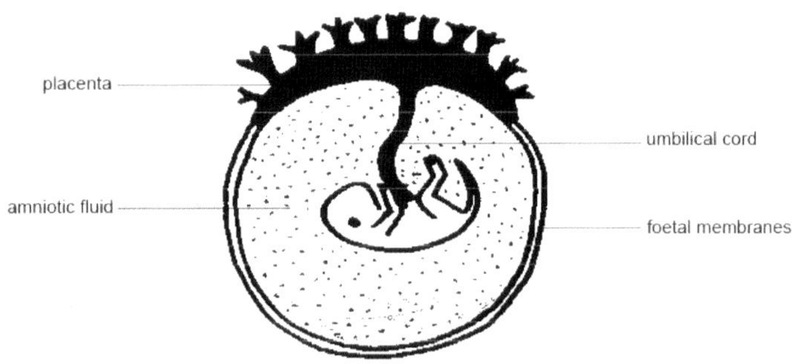

The foetus and placenta

The foetus is attached to the placenta by the umbilical cord. It contains arteries that carry blood to the placenta and a vein that returns blood to the foetus. The developing foetus becomes surrounded by membranes. These enclose the amniotic fluid that protects the foetus from knocks and other trauma.

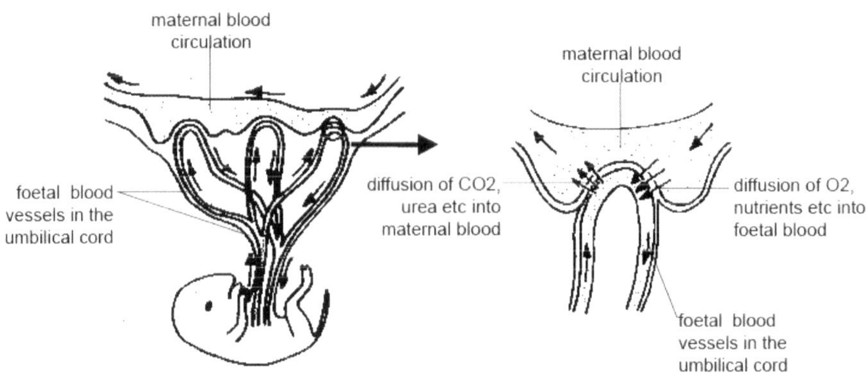

Maternal and foetal blood flow in the placenta

Hormones During Pregnancy

The corpus luteum continues to secrete progesterone and oestrogen during pregnancy. These maintain the lining of the uterus and prepare the mammary glands for milk secretion. Later in the pregnancy the placenta itself takes over the secretion of these hormones.

Chorionic gonadotrophin is another hormone secreted by the placenta and placental membranes. It prevents uterine contractions before labour and prepares the mammary glands for lactation. Towards the end of pregnancy the placenta and ovaries secrete relaxin, a hormone that eases the joint between the two parts of the pelvis and helps dilate the cervix ready for birth.

Pregnancy Testing

The easiest method of pregnancy detection is ultrasound which is noninvasive and very reliable Later in gestation pregnancy can be detected by taking x-rays.

In dogs and cats a blood test can be used to detect the hormone relaxin. In mares and cows palpation of the uterus via the rectum is the classic way to determine pregnancy. It can also be done by detecting the hormones progesterone or equine chorionic gonagotrophin (eCG) in the urine. A new sensitive test measures the amount of the hormone, oestrone sulphate, present in a sample of faeces. The hormone is produced by the foal and placenta, and is only present when there is a living foal.

In most animals, once pregnancy is advanced, there is a window of time during which an experienced veterinarian can determine pregnancy by feeling the abdomen.

Gestation Period

The young of many animals (e.g. pigs, horses and elephants) are born at an advanced state of development, able to stand and even run to escape predators soon after they are born. These animals have a relatively long gestation period that varies with their size e.g. from 114 days in the pig to 640 days in the elephant.

In contrast, cats, dogs, mice, rabbits and higher primates are relatively immature when born and totally dependent on their parents for survival. Their gestation period is shorter and varies from 25 days in the mouse to 31 days in rabbits and 258 days in the gorilla.

The babies of marsupials are born at an extremely immature stage and migrate to the pouch where they attach to a teat to complete their development. Kangaroo joeys, for example, are born 33 days after conception and opossums after only 8 days.

Birth

Signs of Imminent Birth

As the pregnancy continues, the mammary glands enlarge and may secrete a milky substance a few days before birth occurs. The vulva may swell and produce thick mucus and there is sometimes a visible change in the position of the foetus. Just before birth the mother often becomes restless, lying down and getting up frequently. Many animals seek a secluded place where they may build a nest in which to give birth.

Labour

Labour involves waves of uterine contractions that press the foetus against the cervix causing it to dilate. The foetus is then pushed through the cervix and along the vagina before being delivered. In the final stage of labour the placenta or "afterbirth" is expelled.

Adaptations of the Foetus to Life Outside the Uterus

The foetus grows in the watery, protected environment of the uterus where the mother supplies oxygen and nutrients, and waste products pass to her blood circulation for excretion. Once the baby animal is born it must start to breathe for itself, digest food and excrete its own waste. To allow these functions to occur blood is re-routed to the lungs and the glands associated with the gut start to secrete. Note that newborn animals can not control their own body temperature. They need to be kept warm by the mother, littermates and insulating nest materials.

Milk Production

Cows, manatees and primates have two mammary glands but animals like pigs that give birth to large litters may have as many as 12 pairs. Ducts from the gland lead to a nipple or teat and there may be a sinus where the milk collects before being suckled.

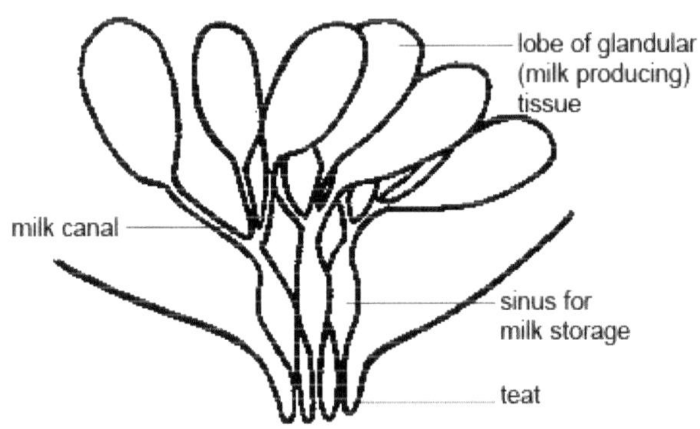

A mammary gland

The hormones oestrogen and progesterone stimulate the mammary glands to develop and prolactin promotes the secretion of the milk. Oxytocin from the pituitary gland releases the milk when the baby suckles. The first milk is called colostrum. It is a rich in nutrients and contains protective antibodies from the mother. Milk contains fat, protein and milk sugar as well as vitamins and most minerals although it contains little iron. Its actual composition varies widely from species to species. For example whale's and seal's milk has twelve times more fat and four times more protein than cow's milk. Cow's milk has far less protein in it than cat's or dog's milk. This is why orphan kittens and puppies cannot be fed cow's milk.

Gonads, Associated Structures, and Products

Although asexual reproduction occurs in many invertebrate species, most reproduce sexually. The basic unit of sexual reproduction is a gamete (sperm or egg), produced by specialized tissues or organs called gonads. Sexual reproduction does not necessarily imply copulation or even a union of gametes. As might be expected of such a large and diverse group as the invertebrates, many variations have evolved to ensure survival of species. In many lower invertebrates, gonads are temporary organs; in higher forms, however, they are permanent. Some invertebrates have coexistent female and male gonads; in others the same gonad produces both sperm and eggs. Animals in which both sperm and eggs are produced by the same individual (hermaphroditism) are termed monoecious. In dioecious species, the sexes are separate. Generally, the male gonads ripen first in hermaphroditic animals (protandry); this tends to ensure cross-fertilization. Self-fertilization is normal, however, in many species, and some species undergo sex reversal.

Animal Breeding

Animal breeding is the selective mating of animals to increase the possibility of obtaining desired traits in the offspring. It has been performed with most domesticated animals, especially cats and dogs, but its main use has been to breed better agricultural stock. The more modern techniques involve a wide variety of laboratory methods, including the modification of embryos, sex selection, and genetic engineering . These procedures are beginning to supplant traditional breeding methods, which focus on selectively combining and isolating livestock strains. In general, the most effective strategy for isolating traits is by selective inbreeding; but different strains are sometimes crossed to take advantage of hybrid vigor and to forestall the negative results of inbreeding, which include reduced fertility, low immunity, and the development of genetic abnormalities.

The Genetic Basis of Animal Breeding

Breeders engage in genetic "experiments" each time they plan a mating. The type of mating selected depends on the goals. To some breeders, determining which traits will appear in the offspring of a mating is like rolling the dice—a combination of luck and chance. For others, producing certain traits involves more skill than luck—the result of careful study and planning. Breeders have to understand how to manipulate genes within their breeding stock to produce the kinds of dogs they want. They have to first understand dogs as a species , then dogs as genetic individuals.

Once the optimal environment for raising an animal to maturity has been established (i.e., the proper nutrition and care has been determined) the only way to manipulate an animal's potential is to manipulate its genetic information. In general, the genetic information of animals is both diverse and uniform: diverse, in the sense that a population will contain many different forms of the same gene (for instance, the human population has 300 different forms of the protein hemoglobin); and uniform, in the sense that there is a basic physical expression of the genetic information that makes, for instance, most goats look similar to each other.

In order to properly understand the basis of animal breeding, it is important to distinguish between genotype and phenotype . Genotype refers to the information contained in an animal's DNA, or genetic material. An animal's phenotype is the physical expression of its genotype. Although every creature is born with a fixed genotype, the phenotype is a variable influenced by many factors in the animal's environment and development. For example, two cows with identical genotypes could develop quite different phenotypes if raised in different environments and fed different foods.

The close association of environment with the expression of the genetic information makes animal breeding a challenging endeavor, because the physical traits a breeder desires to selectively breed for cannot always be attributed entirely to the animal's genes. Moreover, most traits are due not just to one or two genes, but to the complex interplay of many different genes.

DNA consists of a set of chromosomes; the number of chromosomes varies between species (humans, for example, have 46 chromosomes). Mammals (and indeed most creatures) have two copies of each chromosome in the DNA (this is called diploidy). This means there are two copies of the same gene in an animal's DNA. Sometimes each of these will be partially expressed. For example, in a person having one copy of a gene that codes for normal hemoglobin and one coding for sickle-cell hemoglobin, about half of the hemoglobin will be normal and the other half will be sickle-cell. In other cases, only one of the genes can be expressed in the animal's phenotype. The gene expressed is called dominant, and the gene that is not expressed is called recessive. For instance, a human being could have two copies of the gene coding for eye color ; one of them could code for blue, one for brown. The gene coding for brown eyes would be dominant, and the individual's eyes would be brown. But the blue-eyes gene would still exist, and could be passed on to the person's children.

Most of the traits an animal breeder might wish to select will be recessive, for the obvious reason that if the gene were always expressed in the animals, there would be no need to breed for it. If a gene is completely recessive, the animal will need to have two copies of the same gene for it to be expressed (in other words, the animal is homozygous for that particular gene). For this reason, animal breeding is usually most successful when animals are selectively inbred. If a bull has two copies of a gene for a desirable recessive trait, it will pass one copy of this gene to each of its offspring. The other copy of the gene will come from the cow, and assuming it will be normal, none of the offspring will show the desirable trait in their phenotype. However, each of the offspring will have a copy of the recessive gene. If they are then bred with each other, some of their offspring will have two copies of the recessive gene. If two animals with two copies of the recessive gene are bred with each other, all of their offspring will have the desired trait.

There are disadvantages to this method, although it is extremely effective. One of these is that for animal breeding to be performed productively, a number of animals must be involved in the process. Another problem is that undesirable traits can also mistakenly be selected for. For this reason, too much inbreeding will produce sickly or unproductive stock, and at times it is useful to breed two entirely different strains with each other. The resulting offspring are usually extremely healthy; this is referred to as "hybrid vigor." Usually hybrid vigor is only expressed for a generation or two, but crossbreeding is still a very effective means to combat some of the disadvantages of inbreeding.

Another practical disadvantage to selective inbreeding is that the DNA of the parents is altered during the production of eggs and sperm. In order to make eggs and sperm, which are called gametes, a special kind of cell division occurs called meiosis , in which cells divide so that each one has half the normal number of chromosomes (in humans, each sperm and egg contains 23 chromosomes). Before this division occurs, the two pairs of chromosomes wrap around each other, and a phenomenon known as crossing over takes place in which sections of one chromosome will be exchanged with sections of the other chromosome so that new combinations are generated. The problem with crossing over is that some unexpected results can occur. For instance, the offspring of a bull homozygous for two recessive but desirable traits and a cow with "normal" genes will all have one copy of each recessive gene. But when these offspring produce gametes, one recessive gene may migrate to a different chromosome, so that the two traits no longer appear in one gamete. Since most genes work in complicity with others to produce a certain trait, this can make the process of animal breeding very slow, and it requires many generations before the desired traits are obtained—if ever.

Modern Methods in Biotechnology

Although earlier animal breeders had to confine themselves to choosing which of their animals should mate, modern technological advances have altered the face of animal breeding, making it both more selective and more effective. Techniques like genetic engineering , embryo manipulation, artificial insemination, and cloning are becoming more and more refined. Some, like artificial insemination and the manipulation of embryos to produce twins, are now used habitually. Others, such as genetic engineering and cloning, are the subject of intense research and will probably have a great impact on future animal breeding programs.

Artificial Insemination

Artificial insemination is the artificial introduction of semen from a male with desirable traits into females of the species to produce pregnancy, and is useful because a far larger number of offspring can be produced than would be possible if the animals were traditionally bred. Because of this, the value of the male as breeding stock can be determined much more rapidly, and the use of many different females will permit a more accurate evaluation of the hereditability of the desirable traits. In addition, if the traits produced in the offspring do prove to be advantageous, it is easier to disperse them within an animal population in this fashion, as there is a larger breeding stock available. One reason artificial insemination has been an extremely important tool is that it allowed new strains of superior stock to be introduced into a supply of animals in an economically feasible fashion.

The process of artificial insemination requires several steps. Semen must be obtained and effectively diluted, so that the largest number of females can be inseminated consistent with a high probability of pregnancy. The semen must be properly stored so that it remains viable. The females must be tested before the sample is introduced to ensure they are fertile, and following the procedure, they must be tested for pregnancy to determine its success. All these factors make artificial insemination more expensive and more difficult than traditional breeding methods, but the processes have been improved and refined so that the economic advantages far outweigh the procedural disadvantages, and artificial insemination is the most widely applied breeding technique.

Embryo Manipulation

In order to understand the techniques of embryo manipulation, it is important to understand the early stages of reproduction. When the egg and sperm unite to form a zygote, each of the parents supply the zygote with half of the chromosomes necessary for a full set. The zygote, which is a single cell , then begins to reproduce itself by the cellular division process called mitosis , in which each chromosome is duplicated before separation so that each new cell has a full set of chromosomes. This is called the morula stage, and the new cells are called blastomeres. When enough cells have been produced (the number varies from species to species), cell differentiation begins to take place. The first differentiation appears to be when the blastocyst is formed, which is an almost hollow sphere with a cluster of cells inside; and the differentiation appears to be between the cells inside, which become the fetus, and the cells outside, which become the fetal membranes and placenta. However, the process is not entirely understood at the present time and there is some variation between species; so it is difficult to pinpoint the onset of differentiation, which some scientists believe occurs during blastomere division.

During the first stages of cell division, it is possible to separate the blastomeres with the result that each one develops into a separate embryo. Blastomeres with this capability are called totipotent. The purpose of this ability of a single blastomere to produce an entire embryo is probably to safeguard the process of embryo development against the destruction of any of the blastomeres. In theory, it should be possible to produce an entire embryo from each blastomere (and blastomeres are generally totipotent from the four to eight cell stage), but in practice it is usually only possible to produce two embryos. That is why this procedure is generally referred to as embryo splitting rather than cloning, although both terms refer to the same thing (cloning is the production of genetically identical embryos, which is a direct result of embryo splitting).

Interestingly enough, although the embryos produced from separated blastomeres usually have fewer cells than a normal embryo, the resulting offspring fall within the normal range of size for the species.

It is also possible to divide an embryo at other stages of development. For instance, the time at which embryo division is most successful is after the blastocyst has formed. Great care must be taken when dividing a blastocyst, since differentiation has already occurred to some extent, and it is necessary to halve the blastocyst very precisely.

Another interesting embryonic manipulation is the creation of chimaeras. These are formed by uniting two different gametes, so that the embryo has two distinct cell lineages. Chimaeras do not combine the genetic information of both lineages in each cell. Instead, they are a patchwork of cells containing one lineage or the other. For this reason, the offspring of chimaeras are from one distinct genotype or the other, but not from both. Thus chimaeras are not useful for creating new animal populations beyond the first generation. However, they are extremely useful in other contexts. For instance, while embryo division as described above is limited in the number of viable embryos that can be produced, chimaeras can be used to increase the number. After the blastomeres are separated, they can be combined with blastomeres of a different genetic lineage. It has been found that with the additional tissue , the survival rate of the new embryos is more favorable. For some reason only a small percentage of the resulting embryos are chimaeric; this is thought to be because only one cell lineage develops into the cells inside the blastocyst, while the other lineage forms extra-embryonic tissue. It is believed that the more advanced cells are more likely to form the inner cells.

Another application of chimaeras could be for breeding endangered species . Because of the different biochemical environments in the uterus, and the different regulatory mechanisms for fetal development, only very closely related species are able to bear each other's embryos to term. For example, when a goat is implanted with a sheep embryo or the other way around, the embryo is unable to develop properly. This problem can perhaps be surmounted by creating chimaeras in which the placenta stems from the cell lineage of the host species. The immune system of an animal attacks tissue it recognizes as "non-self," but it is possible that the mature chimaeras would be compatible with both the host species and the target species, so that it could bear either embryo to term. This has already proved to be true in studies with mice.

A further technique being developed to manipulate embryos involves the creation of uniparental embryos and same-sex matings. In the former case, the cell from a single gamete is made to go through mitosis, so that the resulting cell is completely homozygous. In the latter case, the DNA from two females (parthogenesis) or two males (androgenesis) is combined to form cells that have only female- or male-derived DNA. These zygotes cannot be developed into live animals, as genetic information from male and female derived DNA is necessary for embryonic development. However, these cells can be used to generate chimaeras. In the case of parthogenetic cells, these chimaeras produce viable gametes. The androgenetic cells do not become incorporated in the embryo; they are used to form extra-embryonic tissue, and so no gametes are recovered.

Aside from these more ambitious embryo manipulation endeavors, multiple ovulation and embryo transfer (MOET) could soon become a useful tool. MOET is the production of multiple embryos from a female with desirable traits, which are then implanted in the wombs of other females of the same species. This circumvents the disadvantages of breeding from a female line (which are that a female can only produce a limited number of offspring due to the time investment and physical rigors of pregnancy). At the present time, MOET is still too expensive for commercial application, but is being applied experimentally.

Genetic Engineering

Genetic engineering is being implemented to create animals that have had a new gene inserted directly into their DNA. These animals are called transgenic. The procedure involves microinjection of the desired gene into the nucleus of fertilized eggs. It has been found that in many cases, but with varying rates of success, the new gene is reproduced in all developing cells, and the gene can be transcribed (which means the information contained in the gene can be read and utilized by the cell). This is a startling breakthrough in animal breeding endeavors, because it means a specific trait can be incorporated into a population in a single generation, rather than the several generations this takes when conventional breeding techniques are used.

However, there are some serious limitations to the procedure. The first of these has to do with the manner in which many genes work together to produce most traits. In fact, there are very few traits a breeder would like to include in an animal population that involve only one or two genes. Although it might some day be possible to incorporate any number of genes into an embryo's DNA, the complex interplay of genes is not understood very well, and the process of identifying all of the genes related to a desired trait is costly and time-consuming.

Another problem in the production of transgenic animals is that they pass their modified DNA on to their offspring with varying success rates and unpredictable results. In some cases, the new gene is present in the offspring but it is not utilized. The new gene may also be altered or rearranged in some way, probably during the process of gamete production.

These factors have made it difficult to successfully produce a transgenic strain of animals. However, with further research into the mechanism by which the gene is incorporated into the genome , and by successfully mapping the target animal genome and identifying the genes responsible for various traits, genetic engineering will no doubt become a major tool for improving animal strains.

Sex Selection

It would be extremely useful if a breeder were able to predetermine the sex of each embryo produced, because in many cases one sex is preferred. For instance, in a herd of dairy cows or a flock of laying hens, females are the only commercially useful sex. When the owner of a dairy herd has inseminated a cow at some expense, this issue becomes morc crucial. In some cases, an animal is being bred specifically for use as breeding stock; in this case, it is far more useful to produce a male that can be bred with multiple females than a female, which can only produce a limited number of offspring.

Whether or not an animal is male or female is determined by its sex chromosomes, which are called X and Y chromosomes. An animal with two X chromosomes will develop into a female, while an animal with one X and one Y chromosome will become a male. In mammals, the sex of the offspring is almost always determined by the male parent, because the female can only donate an X chromosome, and it is the presence or absence of the Y chromosome that causes maleness (this is not true in, for instance, birds ; in that case it is the female who has two different sex chromosomes). The problem in sex selection is to separate the Y-carrying sperm from the X-carrying sperm. Thus far, attempts to do so have been largely unsuccessful or too expensive for commercial application, but the economic advantages make this an area of intense research, and it is quite probable that an efficient and cost-effective method will soon be developed.

Animal Genetics

DNA

Deoxyribonucleic acid (DNA) is the genetic material that contains the instructions in each cell of organisms. DNA determines the genome, and thus the genetic code, which is a blueprint for development of all body organs and structures. The structure of DNA can be visualized as a spiral staircase. The handrails are made up of sugar and phosphate molecules, and the steps are composed of four nitrogenous bases: adenine, thymine, cytosine, and guanine. These bases are paired: adenine is paired with thymine, and cytosine is paired with guanine. The order of these four base pairs is the genetic code that determines the genotype of an individual. The DNA is arranged on chromosomes inside cells, with cells having two methods of dividing and replicating. In mitosis, a cell divides into two daughter cells such that each contains an exact copy of the original cell's chromosomes. In meiosis, a germ cell's chromosomes are duplicated before the cell undergoes two divisions to produce four gametes, or sex cells, each with half (male or female) of the original cell's chromosomes. During the process of fertilization, male and female gametes from different organisms pair their chromosomes to form a zygote, which eventually becomes an adult.

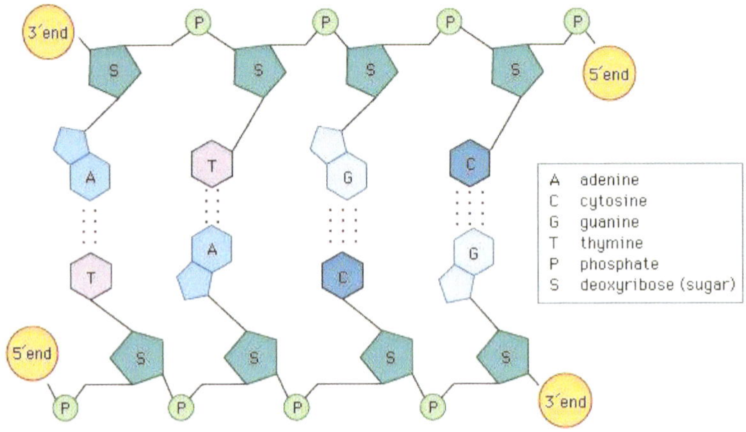

DNA structure, showing the nucleotide bases cytosine (C), thymine (T), adenine (A), and guanine (G) linked to a backbone of alternating phosphate (P) and deoxyribose sugar (S) groups. Two sugar-phosphate chains are paired through hydrogen bonds between A and T and between G and C, thus forming the twin-stranded double helix of the DNA molecule.

Genetic progress in domestic animals has been made using quantitative methods to date. It would be very desirable to know the genes that control the many traits that have economic significance in domestic animals. This should make selection more accurate. Information from sequencing human genes, as well as those of other species, is being used to find chromosomal segments with high probabilities of coding genes in livestock. Another approach is to scan a chromosome segment and look for associations with economic traits. Several quantitative trait loci have been discovered that are or promise to be useful in livestock breeding. For example, an estrogen receptor in pigs is associated with increased litter size—on average, an increase of 0.6 to 2.0 pigs per litter, depending on the genetic background in which the gene is expressed. Other genes have been found that control the secretion of casein in cow's milk. Genes are also known for growth hormone, and many others could be enumerated. With improvements in sequencing DNA, more genes will be discovered that affect economic traits—genes that will need to be tested in different genetic backgrounds and environments before they can be commercialized.

It is now much less expensive to sequence DNA, which has led to new methods of evaluating animals using large segments of 30,000–50,000 bases. With the use of these large segments of DNA, animals are evaluated without looking for markers for individual traits. This is intuitively an appealing approach because much more of the DNA can be evaluated; perhaps in the future the entire genome can be used to evaluate animals.

This method of selection, called genomic selection, is now being applied to dairy cattle, but results are not yet available.

Immunogenetics

The connection between an organism's genetic makeup and its immune system, as well as applications of that knowledge, form the young science of immunogenetics. In particular, producers must control diseases in their livestock if they are going to be profitable. While vaccines, hygiene, and other therapeutic methods control most diseases, vaccines are expensive and none of these methods is completely effective. However, there is evidence from experiments and field data of some degree of genetic control over the immune system in humans and animals. For example, bovine leukocyte adhesion deficiency (BLAD) is a hereditary disease that was discovered in Holstein calves in the 1980s. The presence of the BLAD gene leads to high rates of bacterial infections, pneumonia, diarrhea, and typically death by age four months in cattle, and those that survive their youth have stunted growth and continued susceptibility to infections. It was soon found that these calves carried two copies of a recessive gene that was present in nearly 25 percent of Holstein bulls. Cattle with only one copy of the gene, or carriers, had normal growth patterns and immune systems. Holstein bulls are now routinely tested for the BLAD gene before being used for artificial insemination. With a high percentage of Holsteins being bred artificially, a potentially major problem has been avoided.

Genetic control of the immune system is based on the DNA of the individuals. Histocompatibility genes that serve several functions are on one area of a chromosome, called the major histocompatibility complex (MHC), which exists in all higher vertebrates. There are large numbers of genes involved in the MHCs of different species. There are more than 60 different alleles at one locus and other loci are multi-allelic. There are also differences among species in the number of genes known. In addition, selection experiments have demonstrated genetic variation between lines selected for high and low response to different antigens. Some vaccinations are more efficacious when the animals have been selected for resistance to the antigen for which they are vaccinated.

Substantial progress has been made in the field of immunogenetics, but limited use has been made of this knowledge. One reason for this is that immune systems have evolved to be generally robust. Changing the frequency of some genes that control immune function may inadvertently change the function of other genes and result in adverse effects. Experiments are now under way to determine whether sires' immune responses can be used to predict the health of their daughters under field conditions. The results indicate that there are differences among sires' daughter groups, but the differences are not large enough to control a high proportion of the variability. The tests used were based primarily on leukocytes, which are the first line of defense when an antigen invades an animal. Application of knowledge in the area of immunogenetics must be used with caution.

It might seem that integrating molecular markers and quantitative methods would be a trivial task. However, the effect of some genes depends on the presence of others, and these interactions need to be considered along with the particular breeding scheme. Furthermore, there are nongenetic influences that may turn genes on and off. Thus, some genes act individually, some genes interact, and the environment has a further impact. Finding how these all affect the phenotypic expression of an organism is complicated. However, this challenge presents an opportunity for future research and for producers.

Many advances in reproductive technologies have been made, though many are too expensive for everyday use. Most of the advanced techniques use artificial insemination, which was developed decades ago, though refinements continue.

Cloning

Cloning, an asexual method of reproduction, produces an individual with the same genetic material (DNA) as another individual. Probably the best-known examples of clones are identical twins, which result when cells in the early development stage separate and develop into different individuals. Though the DNA in cloned individuals is the same, environmental influences may make them differ in phenotype. Thus far, the commercial use of clones has been limited. Cloning can be used to produce clones from a highly productive individual, but the cost would have to be low enough to recover the expense quickly. Animals have been cloned by three processes: embryo splitting, blastomere dispersal, and nuclear transfer. Nuclear transfer is most common and involves enucleating an ovum, or egg, with all the genetic material removed. This material is replaced with a full set of chromosomes from a suitable donor cell, which is microinjected into the enucleated cell. Then the enucleated cell, with the transplanted chromosome, is placed into a recipient female to be carried through gestation.

Determining Sex from Sperm

There is a commercial demand for the ability to predetermine the sex of livestock. For example, a producer may want female calves from the best cows for replacements and male calves for beef production. Dairy producers may want more females for replacing cows or for expansion of their herds. The sex of mammals is determined by the sex chromosomes, or X and Y chromosomes. Animals with two X chromosomes develop into females; animals with an X and a Y chromosome develop into males. Thus, the detection of X and Y chromosomes on sperm has been the focus of research to predetermine the sex of domestic animals.

In one process, sperm is pretreated with a dye that fluoresces when exposed to short wavelength light. The fluorescence is brighter from a sperm bearing X chromosomes, which contain about 4 percent more DNA than the Y chromosome. A stream of dyed sperm is passed through a flow cytometer, a computer determines the degree of fluorescence, and the sperm is separated into different containers. The success rate can be as high as 40 percent. When "sexed sperm" has been used on a commercial basis, though, it has had limited success. The conception rate using sexed sperm is lower in cows, though it is higher in primiparous cows. In addition, sperm are killed in the typing process, and the rate of sexing the sperm is slower than desired. While economical processing of sperm is just getting started, it is expected to become another useful tool in animal agriculture.

References

- Sangwal, Keshra (2007). Additives and crystallization processes: from fundamentals to applications. John Wiley and Sons. p. 212. ISBN 978-0-470-06153-4
- Douglas, Angela E.; Raven, John A. (January 2003). "Genomes at the interface between bacteria and organelles". Philosophical Transactions of the Royal Society B. 358 (1429): 5–17. doi:10.1098/rstb.2002.1188. PMC 1693093. PMID 12594915
- "Evolution and Development" (PDF). Carnegie Institution for Science Department of Embryology. 1 May 2012. p. 38. Archived from the original (PDF) on 2 March 2014. Retrieved 4 March2018
- Brusca, Richard C. (2016). Introduction to the Bilateria and the Phylum Xenacoelomorpha | Triploblasty and Bilateral Symmetry Provide New Avenues for Animal Radiation (PDF). Invertebrates. Sinauer Associates. pp. 345–372. ISBN 978-1605353753
- Quillin, K. J. (May 1998). "Ontogenetic scaling of hydrostatic skeletons: geometric, static stress and dynamic stress scaling of the earthworm lumbricus terrestris". The Journal of Experimental Biology. 201 (12): 1871–83. PMID 9600869
- Minkoff, Eli C. (2008). Barron's EZ-101 Study Keys Series: Biology (2nd, revised ed.). Barron's Educational Series. p. 48. ISBN 978-0-7641-3920-8
- Laumer, Christopher E.; Gruber-Vodicka, Harald; Hadfield, Michael G.; Pearse, Vicki B.; Riesgo, Ana; Marioni, John C.; Giribet, Gonzalo (2018-03-17). "Placozoa and Cnidaria are sister taxa". bioRxiv: 200972. doi:10.1101/200972
- Paps, Jordi; Holland, Peter W. H. (30 April 2018). "Reconstruction of the ancestral metazoan genome reveals an increase in genomic novelty". Nature Communications. 9 (1730 (2018)). Bibcode:2018NatCo...9.1730P. doi:10.1038/s41467-018-04136-5. Retrieved 4 May 2018
- Kim, Chang Bae; Moon, Seung Yeo; Gelder, Stuart R.; Kim, Won (September 1996). "Phylogenetic Relationships of Annelids, Molluscs, and Arthropods Evidenced from Molecules and Morphology". Journal of Molecular Evolution. 43 (3): 207–215. doi:10.1007/PL00006079. PMID 8703086
- Sumich, James L. (2008). Laboratory and Field Investigations in Marine Life. Jones & Bartlett Learning. p. 67. ISBN 978-0-7637-5730-4

2

Animals: Vertebrates and Invertebrates

Vertebrates are animals that have backbones, which are surrounded by cartilage or bone. Invertebrates do not have a spinal cord. The topics elaborated in this chapter will help in providing a better perspective about vertebrate and invertebrate animals. It elucidates the five primary vertebrate groups and seven chief invertebrate groups.

Classification of Animals

Classifying animals basically means dividing them into two main groups - vertebrates and invertebrates. Vertebrates have a backbone, while invertebrates are those which don't. In all, there are more than 800,000 animal species in the Kingdom Animalia and most of them are included in the phylum of Arthropod i.e. invertebrates. Usually, people don't think of earthworms or jellyfish as animals, but actually they are, thereby making the animal kingdom classification extremely huge. Each living organism is classified into Kingdom - Phylum - Class - Order - Family - Genus - Species.

The dragonfly serves as a good example of "bilateral symmetry," which means that the right and left sides are essentially mirror images of each other.

Taxonomists typically divide the animal kingdom into two "subkingdoms," which include the invertebrates (animals without backbones) and vertebrates (animals with backbones). As with the plants, taxonomists turn the subsequent animal groupings and classifications, from phyla through genera, into a churning landscape that is simply a part of the scientific process. Depending on their academic roots and research, they divide and re-divide the animal community in many different ways, frequently regrouping, reclassifying and even re-naming species as they go. Some, called "lumpers," identify species as belonging to the same group even though there may be small differences. Other scientists, called "splitters," identify the same species as belonging in distinct groups because of the same small differences. The lumpers produce a relatively simple taxonomy, the splitters, a far more complex taxonomy.

Vertebrate

The vertebrates are commonly called "animals with backbones," but this is a simplified description of a group of animals who are the most anatomically and functionally diverse on Earth. As with most major groups of animals, their beginnings are not known. However, scientists have constructed a theory of the origins of vertebrates that is generally well accepted by the academic community.

Scientists base their model of a hypothetical vertebrate ancestor on several primitive living vertebrates. At the forefront is *Amphioxus,* whose body shape is close to what scientists believe resembles that of the ancestral vertebrate. Many researchers have successfully examined the body of this small marine animal and it has become a popular organism for study in biology classes.

Based on *Amphioxus,* scientists believe that the first vertebrates had a fishlike body with individual segmented muscles along its entire length. They were small ocean dwellers who lived close to the bottom and used their muscles to contract and move their bodies and tail in a side-to-side motion. This action propelled them through the water. They had no distinct head. The brain of early vertebrates is a somewhat controversial topic among scientists, but most researchers agree that the early forms had a brain that was more complex than the simple brain of *Amphioxus.* However, the brain was not the highly complicated structure we see in most vertebrates today. The early brain probably carried out only the most basic of body and sensory functions. As with many animals with cerebral head ganglia, nerve tracts emerged from the brain and ran along the length of the body. Scientists assume the nerve tracts responded to sensory stimuli. The lateral line system, present in most fish, was most likely present in early vertebrates.

Just as in *Amphioxus,* a dorsal (top of body) hollow nerve cord ran the entire length of the body. This nerve cord was supported by an important evolutionary structure made of cartilage called the notocord. The notocord, and its role in the evolution of vertebrates, is one of the most important characteristics distinguishing vertebrates from nonvertebrates. Although there are many other characteristics that help to classify the group, the notocord is the structure from which the backbone is believed to have evolved and is the structure from which vertebrates get their name.

As vertebrates became more specialized and increased their ability to move and sense their environment, the brain and spine became more complex. In looking at the growth of fetal vertebrates it has been shown that the developing muscles place a strain on the cartilaginous notocord. As the strain on the notocord increases with growth of the muscles, deposits of bone replace the cartilage, giving the rod greater strength. This process of replacement eventually produces the bones known as the vertebrae. Each vertebra in a primitive vertebrate corresponds to an individual set of muscles. This pattern is harder to recognize in more derived vertebrates, like mammals and birds, but it is there nonetheless.

In primitive vertebrates, the mouth is a simple oral opening that leads to the gill slits and digestive system. As the vertebrates continued to evolve, the oral cavity was replaced by a more specialized mouth and gill apparatus. Although sharks and their relatives have a primitive type mouth and gills, bony fishes such as salmon and perch have developed complicated gills with a bony covering called an operculum. Sharks are primitive vertebrates in that they do not develop bony skeletons, but even so, the cartilage structure of vertebrates is easy to see.

As vertebrates become complicated in body structure, the mouth becomes a very characteristic structure. Many have teeth that are actually modified body scales. The increasing specialization of the teeth, such as the pointed, socket-bound teeth of the reptile or the many cusped teeth of the mammals, is a major trait on which groups of vertebrates are identified. Birds have no teeth whatsoever.

Early vertebrates had a simple mouth opening through which they gulped food like a frog or fish. This structure was not only poorly adapted for capturing and holding on to active prey but also prevented the animal from breathing while trying to feed. Hunting and swallowing quick prey, like some flying or hopping insects, was problematic to the ill-equipped vertebrates. As a response to increasingly swift food sources that were adapting to life on land, the vertebrates became swifter and more dangerous. The increased specialization of the mouth proved to be an advantage for the group in capturing food.

As they struggled with larger or stronger prey, it was necessary to bite and hold on to wriggling and unwieldy insects. It was hard for them to breathe and many primitive vertebrates were unable to capture these more agile animals.

A major evolutionary trend in the vertebrates was the development of the secondary palate in the mouth, a platform of bone that separates the nasal cavity from the mouth. Mammals, including humans, have a secondary palate that allows for breathing while feeding. This means that the hunter can bite and hold onto its prey and still breathe. It can chew or tear at its food instead of gulping like a crocodile. Lions are an excellent example of how the secondary palate helps the lion to bite and hold onto its intended victim until it is dead and then tear off portions for eating.

It is difficult to provide a generalized summary of the characters of all vertebrates. The group is extremely diverse and includes fish, sharks and rays, amphibians, reptiles, birds, and mammals. However, there are several characteristics that are common to all vertebrates and four that are completely exclusive to the group.

All vertebrates are bilaterally symmetrical—they have two sides which are identical to each other in one plane only. A vertebrate can be divided down the middle, or sagital plane , to produce two equal halves. However, if it is divided down the side, or transverse plane , the sides will not bc idcntical.

The notocord, or skeletal rod, and the dorsal hollow nerve cord are present in all vertebrates. These two characters are unique to vertebrates.

Other characteristics unique to the vertebrates are pharyngeal (at the sides of the pharynx, or throat) gill slits and a tail behind the anal opening. The presence of the tail may seem an obvious trait, but no other group of animals has a structure that can be identified as an actual tail behind the anal opening. Some insects have bodies that extend beyond the anus, but they do not have tails. Vertebrate tails can move and provide locomotion, or balance and support, as in birds and dinosaurs. In many vertebrate groups, such as monkeys, which use their prehensile tails for swinging through trees, the tail can act as an extra appendage.

The vertebrate circulatory system is always closed, but this is not unique to the group. However, the vertebrate heart is always located in a ventral position and the digestive system is complete. Surprising to many, certain vertebrates, especially extinct forms, have an exoskeleton . Early fishes, called ostracoderms and placoderms, had bony exoskeletons that protected their head and sensory areas.

The vertebrates have become a highly successful group of animals with an interesting and exciting evolutionary story. Because humans are vertebrates, they have a natural and continuing curiosity about their predecessors who, most likely, had their beginnings about 500 million years ago in the seas of Earth.

General Features

Although the vertebral column is perhaps the most obvious vertebrate feature, it was not present in the first vertebrates, which probably had only a notochord. The vertebrate has a distinct head, with a differentiated tubular brain and three pairs of sense organs (nasal, optic, and otic). The body is divided into trunk and tail regions. The presence of pharyngeal slits with gills indicates a relatively high metabolic rate. A well-developed notochord enclosed in perichordal connective tissue, with a tubular spinal cord in a connective tissue canal above it, is flanked by a number of segmented muscle masses. A sensory ganglion develops on the dorsal root of the spinal nerve, and segmental autonomic ganglia grow below the notochord. The trunk region is filled with a large, bilateral body cavity (coelom) with contained viscera, and this coelom extends anteriorly into the visceral arches. A digestive system consists of an esophagus extending from the pharynx to the stomach and a gut from the stomach to the anus. A distinct heart, anteroventral to the liver, is enclosed in a pericardial sac. A basic pattern of closed circulatory vessels is largely preserved in most living forms. Unique, bilateral kidneys lie retroperitoneally (dorsal to the main body cavity) and serve blood maintenance and excretory functions. Reproductive organs are formed from tissue adjacent to the kidneys; this original close association is attested by the tubular connections seen in males of living forms. The ducts of the excretory organs open through the body wall into a cloacal chamber, as does the anus of the digestive tract. Reproductive cells are shed through nearby abdominal pores or through special ducts. A muscular tail continues the axial musculature of the trunk.

Approximately 45,000 living species constitute the vertebrates. Species of several classes are found from the high Arctic or Antarctic to the tropics around the Earth; they are missing only from interior Antarctica and Greenland and from the North Polar ice pack. In size, vertebrates range from minute fishes to elephants and whales (of up to 100 tons), the largest animals ever to have existed. Vertebrates are adapted to life underground, on the surface, and in the air. They feed upon plants, invertebrate animals, and one another. Vertebrate faunas are important to humans for food and recreation.

Reproductive Systems

Nearly all vertebrates undergo sexual reproduction. They produce haploid gametes by meiosis. The smaller, motile gametes are spermatozoa and the larger, non-motile gametes are ova. These fuse by the process of fertilisation to form diploid zygotes, which develop into new individuals.

Inbreeding

During sexual reproduction, mating with a close relative (inbreeding) often leads to inbreeding depression. Inbreeding depression is considered to be largely due to expression of deleterious recessive mutations. The effects of inbreeding have been studied in many vertebrate species.

In several species of fish, inbreeding was found to decrease reproductive success.

Inbreeding was observed to increase juvenile mortality in 11 small animal species.

A common breeding practice for pet dogs is mating between close relatives (e.g. between half- and full siblings). This practice generally has a negative effect on measures of reproductive success, including decreased litter size and puppy survival.

Incestuous matings in birds result in severe fitness costs due to inbreeding depression (e.g. reduction in hatchability of eggs and reduced progeny survival).

Inbreeding Avoidance

As a result of the negative fitness consequences of inbreeding, vertebrate species have evolved mechanisms to avoid inbreeding. Numerous inbreeding avoidance mechanisms operating prior to mating have been described.

Toads and many other amphibians display breeding site fidelity. Individuals that return to natal ponds to breed will likely encounter siblings as potential mates. Although incest is possible, *Bufo americanus* siblings rarely mate. These toads likely recognize and actively avoid close kins as mates. Advertisement vocalizations by males appear to serve as cues by which females recognize their kin.

Inbreeding avoidance mechanisms can also operate subsequent to copulation. In guppies, a post-copulatory mechanism of inbreeding avoidance occurs based on competition between sperm of rival males for achieving fertilization. In competitions between sperm from an unrelated male and from a full sibling male, a significant bias in paternity towards the unrelated male was observed.

When female sand lizards mate with two or more males, sperm competition within the female's reproductive tract may occur. Active selection of sperm by females appears to occur in a manner that enhances female fitness. On the basis of this selective process, the sperm of males that are more distantly related to the female are preferentially used for fertilization, rather than the sperm of close relatives. This preference may enhance the fitness of progeny by reducing inbreeding depression.

Outcrossing

Mating with unrelated or distantly related members of the same species is generally thought to provide the advantage of masking deleterious recessive mutations in progeny. Vertebrates have evolved numerous diverse mechanisms for avoiding close inbreeding and promoting outcrossing.

Outcrossing as a way of avoiding inbreeding depression, has been especially well studied in birds. For instance, inbreeding depression occurs in the great tit when the offspring are produced as a result of a mating between close relatives. In natural populations of the great tit (*Parus major*), inbreeding is avoided by dispersal of individuals from their birthplace, which reduces the chance of mating with a close relative.

The purple-crowned fairywren females paired with related males may undertake extra-pair matings that can reduce the negative effects of inbreeding. However, there are ecological and demographic constraints on extra pair matings. Nevertheless, 46% of broods produced by incestuously paired females contained extra-pair young.

Southern pied babblers (*Turdoides bicolor*) appear to avoid inbreeding in two ways. The first is through dispersal, and the second is by avoiding familiar group members as mates. Although both males and females disperse locally, they move outside the range where genetically related individuals are likely to be encountered. Within their group, individuals only acquire breeding positions when the opposite-sex breeder is unrelated.

Cooperative breeding in birds typically occurs when offspring, usually males, delay dispersal from their natal group in order to remain with the family to help rear younger kin. Female offspring rarely stay at home, dispersing over distances that allow them to breed independently, or to join unrelated groups.

Parthenogenesis

Parthenogenesis is a natural form of reproduction in which growth and development of embryos occur without fertilization.

Reproduction in squamate reptiles is ordinarily sexual, with males having a ZZ pair of sex determining chromosomes, and females a ZW pair. However, various species, including the Colombian Rainbow boa (*Epicrates maurus*), *Agkistrodon contortrix* (copperhead snake) and *Agkistrodon piscivorus* (cotton mouth snake) can also reproduce by facultative parthenogenesis -that is, they are capable of switching from a sexual mode of reproduction to an asexual mode- resulting in production of WW female progeny. The WW females are likely produced by terminal automixis.

Mole salamanders are an ancient (2.4-3.8 million year-old) unisexual vertebrate lineage. In the polyploid unisexual mole salamander females, a premeiotic endomitotic event doubles the number of chromosomes. As a result, the mature eggs produced subsequent to the two meiotic divisions have the same ploidy as the somatic cells of the female salamander. Synapsis and recombination during meiotic prophase I in these unisexual females is thought to ordinarily occur between identical sister chromosomes and occasionally between homologous chromosomes. Thus little, if any, genetic variation is produced. Recombination between homeologous chromosomes occurs only rarely, if at all. Since production of genetic variation is weak, at best, it is unlikely to provide a benefit sufficient to account for the long-term maintenance of meiosis in these organisms. However, meiosis may have been maintained during evolution by the efficient recombinational repair of DNA damages that meiosis provides, an advantage that could be realized at each generation.

Self-fertilization

The mangrove killifish (*Kryptolebias marmoratus*) produces both eggs and sperm by meiosis and routinely reproduces by self-fertilisation. The capacity for selfing in these fishes has apparently persisted for at least several hundred thousand years. Each individual hermaphrodite normally fertilizes itself when an egg and sperm that it has produced by an internal organ unite inside the fish's body. In nature, this mode of reproduction can yield highly homozygous lines composed of individuals so genetically uniform as to be, in effect, identical to one another. Although inbreeding, especially in the extreme form of self-fertilization, is ordinarily regarded as detrimental because it leads to expression of deleterious recessive alleles, self-fertilization does provide the benefit of *fertilization assurance* (reproductive assurance) at each generation.

Classifying a Vertebrate

The vertebrate subkingdom of our deserts includes phyla comprising a surprisingly diversity of mammals, birds, reptiles, amphibians and even fish.

"The Southwest contributes impressively to the continentwide diversity of mammals; native mammal species in southwestern states number about 120 in Texas, 138 in Arizona, 139 in New Mexico, and 163 in California. No other region in the country has so many mammal species," says the United States Geological Survey in its Science for a Changing World Internet site.

The western Bluebird is one of nearly 400 species of birds in the southwest.

The Southwest contributes even more to the diversity of birds. Southeastern Arizona's Chiricahua Mountains bird "checklist presently stands at 375 species [including the splendid Elegant Trogon], not including 18 species still considered hypothetical—about half of all the birds regular occurring on this continent north of Mexico," according to Richard C. Taylor's *Location Checklist to the Birds of the Chiricahua Mountains.* Central New Mexico's Bosque del Apache's checklist "contains 377 species which have been observed on the refuge since 1940," according to the USGS Bird Checklists of the United States.

The nocturnal, smooth-scaled Glossy is one of nearly five dozen species of snakes in the southwest.

As for reptiles and amphibians, more than 150 species "occur in the Southwest; this is partly due to the region's diverse habitats," says the USGS in *Science for a Changing World.* "Compared with the moister habitats of eastern North America, the southwestern amphibian fauna is relatively poor, with only 3 salamander species and about 30 frog and toad species. About 25% of the amphibians, though, are found nowhere else in the United States. About 50 lizard and 56 snake species account for more than 70% of the reptiles and amphibians inhabiting this region; most of these live in the arid and semiarid desert grasslands and shrublands that make up most of the Southwest."

The Southwest's few water sources host a surprising variety of fish, although many of the species are imperiled by human assault on their habitat. According to the USGS, the Colorado River drainage system is home to numerous species including various chubs, the Colorado squawfish, the razorback sucker and the bonytail. The Rio Grande system of the Colorado River drainage system is home to chubs, the silvery minnow and shiners. Desert springs serve as habitat for pupfishes, springfishes and poolfishes.

An icon of the Southwest, the coyote, serves as a guide to how the classification system works for the vertebrates.

The Coyote is called Canis latrans, or barking dog.

At the phylum level, the coyote belongs to the *Chordata,* which is distinguished by a nerve chord that passes through the backbone. At the class level, the animal belongs to the *Mammalia,* the members of which possess fur and nurse their young. At the order level, the coyote, like his fellow *Carnivora,* shares "extreme night vision capabilities and various more broadly defined characteristics including: high intelligence, relatively long maturation period for the young, dental similarities, tendency for complex social organization, and generally bearing 2-3 young in a litter (but they may birth as many as 4-5 at one time)," according to the Tigerhomes.org Internet site. At the family level, our guide and his close relatives, collectively called *Canidae,* typically have elongated faces. They have similar tooth arrangements, or "dental formulas." Unlike, say, a bear, which belongs to another family of mammals, the *Canidae* walk on their toes. At the genus level, the coyote and the related species – including the gray wolf (northern Chihuahuan Desert and the Colorado Plateau), the red wolf (western Piney Woods) and your family dog – belong to the *Canis*. They all bear wolf-like characteristics, and they can interbreed, provided the male and female are of compatible size. (A male chihuahua would experience a difficult time mating with a female gray wolf.)

At the species level, the coyote is called *Canis latrans,* or barking dog, an allusion to the animal's vocal talents. It is about the size and shape of a medium-sized Collie dog, but its tail is round and bushy and is carried straight out below the level of its back. "Coyotes found in low deserts and valleys weigh about 20 pounds, less than half of their mountain kin, who can weigh up to 50 pounds. Desert coyotes are light gray or tan with a black tip on the tail." William H. Burt and Richard P. Grossenheider point out in their *A Field Guide to the Mammals of America North of Mexico,* the coyote's nose "is more pointed and [the] tail is bushier than normal in dogs; [its tail is] *held down* between [its] hind legs *when running.*" In summary, it fits into the Linnaeus classification scheme as follows:

Classification Level	Scientific Name
Kingdom	Animalia
Subkingdom	Vertebrate
Phylum	*Chordata*
Class	*Mammalia*
Order	*Carnivora*
Family	*Canidae*
Genus	*Canis*
Species	*Latrans*
Common Name	Coyote
Scientific Name	*Canis latrans*

The coyote has long played a central part in the folk tales, myths and persona of the peoples of the Southwestern deserts, often taking on the role of clown or adventurer. In stories recorded by Charles F. Lummis in his *Pueblo Indian Folk-Stories,* foolish coyote falls for practical jokes played by ravens, blackbirds and a bear. Compassionate coyote saves an abandoned baby, bestowing it upon a mother antelope that has lost her fawn. Magical coyote jumps through a willow hoop and emerges as a handsome young man.

While fellow *Canis* Mexican gray wolf disappeared (until being reintroduced) from the Southwestern deserts in historic times, clever, adaptable and bold coyote has increased its population and expanded its range from its native desert basins well up into the mountains. Thief coyote can make a nuisance of himself.

Main Vertebrate Group

Mammals

Mammal, (class Mammalia), any member of the group of vertebrate animals in which the young are nourished with milk from special mammary glands of the mother. In addition to these characteristic milk glands, mammals are distinguished by several other unique features. Hair is a typical mammalian feature, although in many whales it has disappeared except in the fetal stage. The mammalian lower jaw is hinged directly to the skull, instead of through a separate bone (the quadrate) as in all other vertebrates. A chain of three tiny bones transmits sound waves across the middle ear. A muscular diaphragm separates the heart and the lungs from the abdominal cavity. Only the left aortic arch persists. (In birds the right aortic arch persists; in reptiles, amphibians, and fishes both arches are retained.) Mature red blood cells (erythrocytes) in all mammals lack a nucleus; all other vertebrates have nucleated red blood cells.

Except for the monotremes (an egg-laying order of mammals comprising echidnas and the duck-billed platypus), all mammals are viviparous—they bear live young. In the placental mammals (which have a placenta to facilitate nutrient and waste exchange between the mother and the developing fetus), the young are carried within the mother's womb, reaching a relatively advanced stage of development before birth. In the marsupials (e.g., kangaroos, opossums, and wallabies), the newborns are incompletely developed at birth and continue to develop outside the womb, attaching themselves to the female's body in the area of her mammary glands. Some marsupials have a pouchlike structure or fold, the marsupium, that shelters the suckling young.

The class Mammalia is worldwide in distribution. It has been said that mammals have a wider distribution and are more adaptable than any other single class of animals, with the exception of certain less-complex forms such as arachnids and insects. This versatility in exploiting Earth is attributed in large part to the ability of mammals to regulate their body temperatures and internal environment both in excessive heat and aridity and in severe cold.

Anatomy and Morphology

Distinguishing Features

Living mammal species can be identified by the presence of sweat glands, including those that are specialized to produce milk to nourish their young. In classifying fossils, however, other features must be used, since soft tissue glands and many other features are not visible in fossils.

Many traits shared by all living mammals appeared among the earliest members of the group:

- Jaw joint - The dentary (the lower jaw bone, which carries the teeth) and the squamosal (a small cranial bone) meet to form the joint. In most gnathostomes, including early therapsids, the joint consists of the articular (a small bone at the back of the lower jaw) and quadrate (a small bone at the back of the upper jaw).
- Middle ear - In crown-group mammals, sound is carried from the eardrum by a chain of three bones, the malleus, the incus and the stapes. Ancestrally, the malleus and the incus are derived from the articular and the quadrate bones that constituted the jaw joint of early therapsids.
- Tooth replacement - Teeth are replaced once or (as in toothed whales and murid rodents) not at all, rather than being replaced continually throughout life.
- Prismatic enamel - The enamel coating on the surface of a tooth consists of prisms, solid, rod-like structures extending from the dentin to the tooth's surface.
- Occipital condyles - Two knobs at the base of the skull fit into the topmost neck vertebra; most other tetrapods, in contrast, have only one such knob.

For the most part, these characteristics were not present in the Triassic ancestors of the mammals. Nearly all mammaliaforms possess an epipubic bone, the exception being modern placentals.

Biological Systems

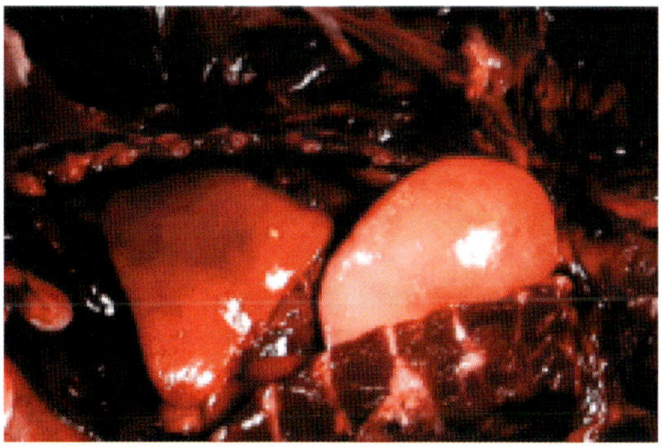

Raccoon lungs

The majority of mammals have seven cervical vertebrae (bones in the neck), including bats, giraffes, whales and humans. The exceptions are the manatee and the two-toed sloth, which have just six, and the three-toed sloth which has nine cervical vertebrae. All mammalian brains possess a neocortex, a brain region unique to mammals. Placental mammals have a corpus callosum, unlike monotremes and marsupials.

The lungs of mammals are spongy and honeycombed. Breathing is mainly achieved with the diaphragm, which divides the thorax from the abdominal cavity, forming a dome convex to the thorax. Contraction of the diaphragm flattens the dome, increasing the volume of the lung cavity. Air enters through the oral and nasal cavities, and travels through the larynx, trachea and bronchi, and expands the alveoli. Relaxing the diaphragm has the opposite effect, decreasing the volume of the lung cavity, causing air to be pushed out of the lungs. During exercise, the abdominal wall contracts, increasing pressure on the diaphragm, which forces air out quicker and more forcefully. The rib cage is able to expand and contract the chest cavity through the action of other respiratory muscles. Consequently, air is sucked into or expelled out of the lungs, always moving down its pressure gradient. This type of lung is known as a bellows lung due to its resemblance to blacksmith bellows.

The mammalian heart has four chambers, two upper atria, the receiving chambers, and two lower ventricles, the discharging chambers. The heart has four valves, which separate its chambers and ensures blood flows in the correct direction through the heart (preventing backflow). After gas exchange in the pulmonary capillaries (blood vessels in the lungs), oxygen-rich blood returns to the left atrium via one of the four pulmonary veins. Blood flows nearly continuously back into the atrium, which acts as the receiving chamber, and from here through an opening into the left ventricle. Most blood flows passively into the heart while both the atria and ventricles are relaxed, but toward the end of the ventricular relaxation period, the left atrium will contract, pumping blood into the ventricle. The heart also requires nutrients and oxygen found in blood like other muscles, and is supplied via coronary arteries.

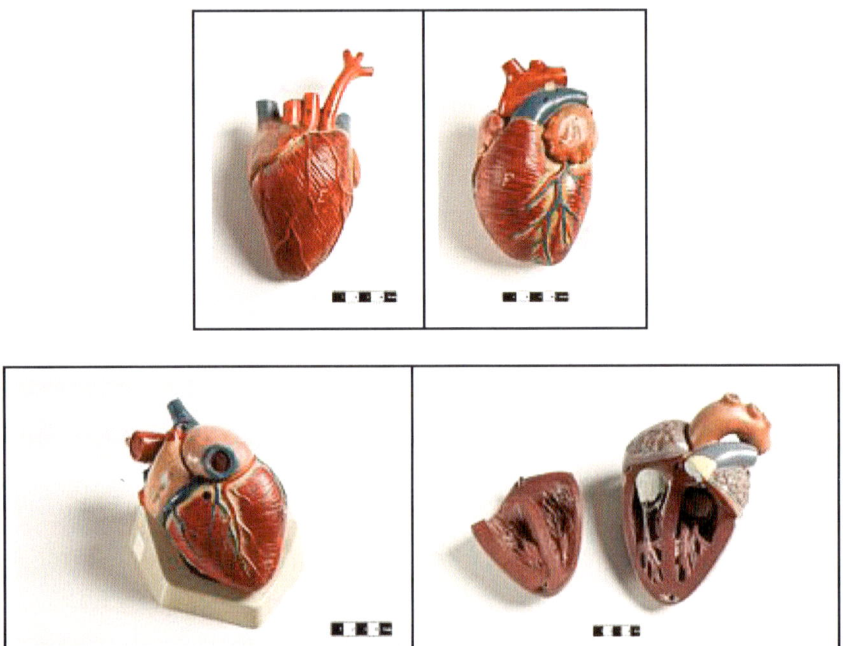

Didactic models of a mammalian heart

The integumentary system is made up of three layers: the outermost epidermis, the dermis and the hypodermis. The epidermis is typically 10 to 30 cells thick; its main function is to provide a waterproof layer. Its outermost cells are constantly lost; its bottommost cells are constantly dividing and pushing upward. The middle layer, the dermis, is 15 to 40 times thicker than the epidermis. The dermis is made up of many components, such as bony structures and blood vessels. The hypodermis is made up of adipose tissue, which stores lipids and provides cushioning and insulation. The thickness of this layer varies widely from species to species; marine mammals require a thick hypodermis (blubber) for insulation, and right whales have the thickest blubber at 20 inches (51 cm). Although other animals have features such as whiskers, feathers, setae, or cilia that superficially resemble it, no animals other than mammals have hair. It is a definitive characteristic of the class. Though some mammals have very little, careful examination reveals the characteristic, often in obscure parts of their bodies.

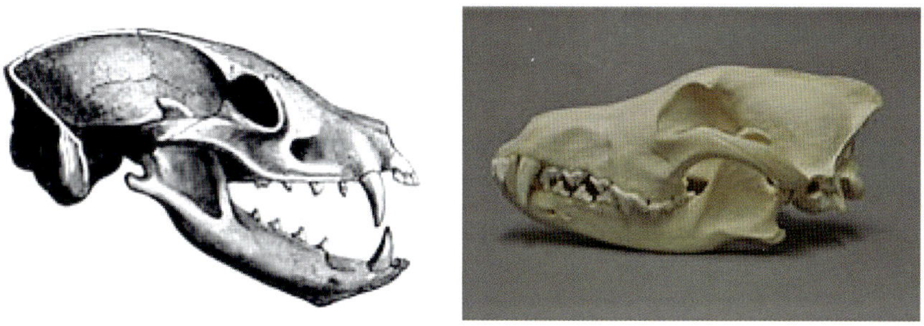

The carnassials (teeth in the very back of the mouth) of the insectivorous aardwolf (left) vs. that of a gray wolf (right) which consumes large vertebrates

Herbivores have developed a diverse range of physical structures to facilitate the consumption of plant material. To break up intact plant tissues, mammals have developed teeth structures that reflect their feeding preferences. For instance, frugivores (animals that feed primarily on fruit) and herbivores that feed on soft foliage have low-crowned teeth specialized for grinding foliage and seeds. Grazing animals that tend to eat hard, silica-rich grasses, have high-crowned teeth, which are capable of grinding tough plant tissues and do not wear down as quickly as low-crowned teeth. Most carnivorous mammals have carnassialiforme teeth (of varying length depending on diet), long canines and similar tooth replacement patterns.

The stomach of Artiodactyls is divided into four sections: the rumen, the reticulum, the omasum and the abomasum (only ruminants have a rumen). After the plant material is consumed, it is mixed with saliva in the rumen and reticulum and separates into solid and liquid material. The solids lump together to form a bolus (or cud), and is regurgitated. When the bolus enters the mouth, the fluid is squeezed out with the tongue and swallowed again. Ingested food passes to the rumen and reticulum where cellulytic microbes (bacteria, protozoa and fungi) produce cellulase, which is needed to break down the cellulose in plants. Perissodactyls, in contrast to the ruminants, store digested food that has left the stomach in an enlarged cecum, where it is fermented by bacteria. Carnivora have a simple stomach adapted to digest primarily meat, as compared to the elaborate digestive systems of herbivorous animals, which are necessary to break down tough, complex plant fibers. The caecum is either absent or short and simple, and the large intestine is not sacculated or much wider than the small intestine.

Bovine kidney

The mammalian excretory system involves many components. Like most other land animals, mammals are ureotelic, and convert ammonia into urea, which is done by the liver as part of the urea cycle. Bilirubin, a waste product derived from blood cells, is passed through bile and urine with the help of enzymes excreted by the liver. The passing of bilirubin via bile through the intestinal tract gives mammalian feces a distinctive brown coloration. Distinctive features of the mammalian kidney include the presence of the renal pelvis and renal pyramids, and of a clearly distinguishable cortex and medulla, which is due to the presence of elongated loops of Henle. Only the mammalian kidney has a bean shape, although there are some exceptions, such as the multilobed reniculate kidneys of pinnipeds, cetaceans and bears. Most adult placental mammals have no remaining trace of the cloaca. In the embryo, the embryonic cloaca divides into a posterior region that becomes part of the anus, and an anterior region that has different fates depending on the sex of the individual: in females, it develops into the vestibule that receives the urethra and vagina, while in males it forms the entirety of the penile urethra. However, the tenrecs, golden moles, and some shrews retain a cloaca as adults. In marsupials, the genital tract is separate from the anus, but a trace of the original cloaca does remain externally. Monotremes, which translates from Greek into “single hole”, have a true cloaca.

Sound Production

As in all other tetrapods, mammals have a larynx that can quickly open and close to produce sounds, and a supralaryngeal vocal tract which filters this sound. The lungs and surrounding musculature provide the air stream and pressure required to phonate. The larynx controls the pitch and volume of sound, but the strength the lungs exert to exhale also contributes to volume. More primitive mammals, such as the echidna, can only hiss, as sound is achieved solely through exhaling through a partially closed larynx. Other mammals phonate using vocal folds, as opposed to the vocal cords seen in birds and reptiles. The movement or tenseness of the vocal folds can result in many sounds such as purring and screaming. Mammals can change the position of the larynx, allowing them to breathe through the nose while swallowing through the mouth, and to form both oral and nasal sounds; nasal sounds, such as a dog whine, are generally soft sounds, and oral sounds, such as a dog bark, are generally loud.

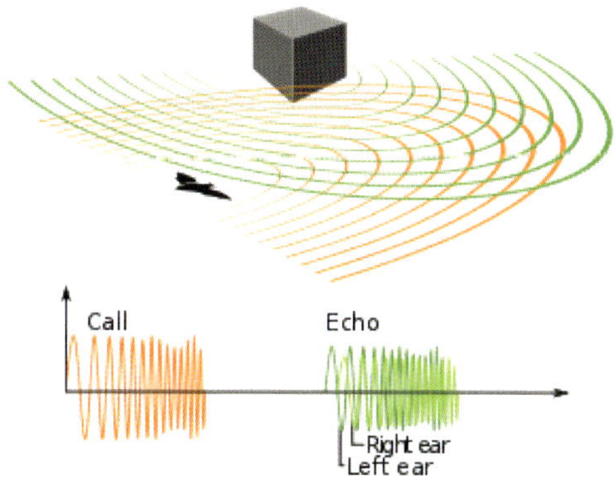

A diagram of ultrasonic signals emitted by a bat, and the echo from a nearby object

Some mammals have a large larynx and thus a low-pitched voice, namely the hammer-headed bat (*Hypsignathus monstrosus*) where the larynx can take up the entirety of the thoracic cavity while pushing the lungs, heart, and trachea into the abdomen. Large vocal pads can also lower the pitch, as in the low-pitched roars of big cats. The production of infrasound is possible in some mammals such as the African elephant (*Loxodonta* spp.) and baleen whales. Small mammals with small larynxes have the ability to produced ultrasound, which can be detected by modifications to the middle ear and cochlea. Ultrasound is inaudible to birds and reptiles, which might have been important during the Mesozoic, when birds and reptiles were the dominant predators. This private channel is used by some rodents in, for example, mother-to-pup communication, and by bats when echolocating. Toothed whales also use echolocation, but, as opposed to the vocal membrane that extends upward from the vocal folds, they have a melon to manipulate sounds. Some mammals, namely the primates, have air sacs attached to the larynx, which may function to lower the resonances or increase the volume of sound.

The vocal production system is controlled by the cranial nerve nuclei in the brain, and supplied by the recurrent laryngeal nerve and the superior laryngeal nerve, branches of the vagus nerve. The vocal tract is supplied by the hypoglossal nerve and facial nerves. Electrical stimulation of the periaqueductal gray (PEG) region of the mammalian midbrain elicit vocalizations. The ability to learn new vocalizations is only exemplified in humans, seals, cetaceans, elephants and possibly bats; in humans, this is the result of a direct connection between the motor cortex, which controls movement, and the motor neurons in the spinal cord.

Fur

Porcupines use their spines for defense.

The fur of mammals has many uses protection, sensory purposes, waterproofing, and camouflage, with the primary usage being thermoregulation. The types of hair include definitive, which may be shed after reaching a certain length; vibrissae, which are sensory hairs and are most commonly whiskers; pelage, which consists of guard hairs, under-fur, and awn hair; spines, which are a type of stiff guard hair used for defense in, for example, porcupines; bristles, which are long hairs usually used in visual signals, such as the mane of a lion; velli, often called "down fur," which insulates newborn mammals; and wool which is long, soft and often curly. Hair length is negligible in thermoregulation, as some tropical mammals, such as sloths, have the same length of fur length as some arctic mammals but with less insulation; and, conversely, other tropical mammals with short hair have the same insulating value as arctic mammals. The denseness of fur can increase an animal's insulation value, and arctic mammals especially have dense fur; for example, the musk ox has guard hairs measuring 30 cm (12 in) as well as a dense underfur, which forms an airtight coat, allowing them to survive in temperatures of −40 °C (−40 °F). Some desert mammals, such as camels, use dense fur to prevent solar heat from reaching their skin, allowing the animal to stay cool; a camel's fur may reach 70 °C (158 °F) in the summer, but the skin stays at 40 °C (104 °F). Aquatic mammals, conversely, trap air in their fur to conserve heat by keeping the skin dry.

A leopard's disruptively colored coat provides camouflage for this ambush predator.

Mammalian coats are colored for a variety of reasons, the major selective pressures including camouflage, sexual selection, communication and physiological processes such as temperature regulation. Camouflage is a powerful influence in a large number of mammals, as it helps to conceal individuals from predators or prey. Aposematism, warning off possible predators, is the most likely explanation of the black-and-white pelage of many mammals which are able to defend themselves, such as in the foul-smelling skunk and the powerful and aggressive honey badger. In arctic and subarctic mammals such as the arctic fox (*Alopex lagopus*), collared lemming (*Dicrostonyx groenlandicus*), stoat (*Mustela erminea*), and snowshoe hare (*Lepus americanus*), seasonal color change between brown in summer and white in winter is driven largely by camouflage. Differences in female and male coat color may indicate nutrition and hormone levels, important in mate selection. Some arboreal mammals, notably primates and marsupials, have shades of violet, green, or blue skin on parts of their bodies, indicating some distinct advantage in their largely arboreal habitat due to convergent evolution. The green coloration of sloths, however, is the result of a symbiotic relationship with algae. Coat color is sometimes sexually dimorphic, as in many primate species. Coat color may influence the ability to retain heat, depending on how much light is reflected. Mammals with a darker colored coat can absorb more heat from solar radiation, and stay warmer, and some smaller mammals, such as voles, have darker fur in the winter. The white, pigmentless fur of arctic mammals, such as the polar bear, may reflect more solar radiation directly onto the skin.

Reproductive System

Goat kids stay with their mother until they are weaned.

In male placentals, the penis is used both for urination and copulation. Depending on the species, an erection may be fueled by blood flow into vascular, spongy tissue or by muscular action. A penis may be contained in a sheath when not erect, and some placentals also have a penis bone (baculum). Marsupials typically have forked penises while the monotreme penis generally has four heads with only two functioning. The testes of most mammals descend into the scrotum which is typically posterior to the penis but is often anterior in marsupials. Female mammals generally have a clitoris, labia majora and labia minora on the outside, while the internal system contains paired oviducts, 1-2 uteri, 1-2 cervices and a vagina. Marsupials have two lateral vaginas and a medial vagina. The "vagina" of monotremes is better understood as a "urogenital sinus". The uterine systems of placental mammals can vary between a duplex, were there are two uteri and cervices which open into the vagina, a bipartite, were two uterine horns have a single cervix that connects to the vagina, a bicornuate, which consists where two uterine horns that are connected distally but separate medially creating a Y-shape, and a simplex, which has a single uterus.

Most mammals are viviparous, giving birth to live young. However, the five species of monotreme, the platypus and the four species of echidna, lay eggs. The monotremes have a sex determination system different from that of most other mammals. In particular, the sex chromosomes of a platypus are more like those of a chicken than those of a therian mammal.

Viviparous mammals are in the subclass Theria; those living today are in the marsupial and placental infraclasses. Marsupials have a short gestation period, typically shorter than its estrous cycle and gives birth to an undeveloped newborn that then undergoes further development; in many species, this takes place within a pouch-like sac, the marsupium, located in the front of the mother's abdomen. This is the plesiomorphic condition among viviparous mammals; the presence of epipubic bones in all non-placental mammals prevents the expansion of the torso needed for full pregnancy. Even non-placental eutherians probably reproduced this way. The placentals give birth to relatively complete and developed young, usually after long gestation periods. They get their name from the placenta, which connects the developing fetus to the uterine wall to allow nutrient uptake.

The mammary glands of mammals are specialized to produce milk, the primary source of nutrition for newborns. The monotremes branched early from other mammals and do not have the nipples seen in most mammals, but they do have mammary glands. The young lick the milk from a mammary patch on the mother's belly.

Endothermy

Nearly all mammals are endothermic ("warm-blooded"). Most mammals also have hair to help keep them warm. Like birds, mammals can forage or hunt in weather and climates too cold for ectothermic ("cold-blooded") reptiles and insects. Endothermy requires plenty of food energy, so mammals eat more food per unit of body weight than most reptiles. Small insectivorous mammals eat prodigious amounts for their size. A rare exception, the naked mole-rat produces little metabolic heat, so it is considered an operational poikilotherm. Birds are also endothermic, so endothermy is not unique to mammals.

Species Lifespan

Among mammals, species maximum lifespan varies significantly (for example the shrew has a lifespan of two years, whereas the oldest bowhead whale is recorded to be 211 years). Although the underlying basis for these lifespan differences is still uncertain, numerous studies indicate that the ability to repair DNA damages is an important determinant of mammalian lifespan. In a 1974 study by Hart and Setlow, it was found that DNA excision repair capability increased systematically with species lifespan among seven mammalian species. Species lifespan was observed to be robustly correlated with the capacity to recognize DNA double-strand breaks as well as the level of the DNA repair protein Ku80. In a study of the cells from sixteen mammalian species, genes employed in DNA repair were found to be up-regulated in the longer-lived species. The cellular level of the DNA repair enzyme poly ADP ribose polymerase was found to correlate with species lifespan in a study of 13 mammalian species. Three additional studies of a variety of mammalian species also reported a correlation between species lifespan and DNA repair capability.

Locomotion

Terrestrial

Running gait

Most vertebrates—the amphibians, the reptiles and some mammals such as humans and bears—are plantigrade, walking on the whole of the underside of the foot. Many mammals, such as cats and dogs, are digitigrade, walking on their toes, the greater stride length allowing more speed. Digitigrade mammals are also often adept at quiet movement. Some animals such as horses are unguligrade, walking on the tips of their toes. This even further increases their stride length and thus their speed. A few mammals, namely the great apes, are also known to walk on their knuckles, at least for their front legs. Giant anteaters and platypuses are also knuckle-walkers. Some mammals are bipeds, using only two limbs for locomotion, which can be seen in, for example, humans and the great apes. Bipedal species have a larger field of vision than quadrupeds, conserve more energy and have the ability to manipulate objects with their hands, which aids in foraging. Instead of walking, some bipeds hop, such as kangaroos and kangaroo rats.

Animals will use different gaits for different speeds, terrain and situations. For example, horses show four natural gaits, the slowest horse gait is the walk, then there are three faster gaits which, from slowest to fastest, are the trot, the canter and the gallop. Animals may also have unusual gaits that are used occasionally, such as for moving sideways or backwards. For example, the main human gaits are bipedal walking and running, but they employ many other gaits occasionally, including a four-legged crawl in tight spaces. Mammals show a vast range of gaits, the order that they place and lift their appendages in locomotion. Gaits can be grouped into categories according to their patterns of support sequence. For quadrupeds, there are three main categories: walking gaits, running gaits and leaping gaits. Walking is the most common gait, where some feet are on the ground at any given time, and found in almost all legged animals. Running is considered to occur when at some points in the stride all feet are off the ground in a moment of suspension.

Arboreal

Gibbons are very good brachiators because their elongated limbs enable them to easily swing and grasp on to branches.

Arboreal animals frequently have elongated limbs that help them cross gaps, reach fruit or other resources, test the firmness of support ahead and, in some cases, to brachiate (swing between trees). Many arboreal species, such as tree porcupines, silky anteaters, spider monkeys, and possums, use prehensile tails to grasp branches. In the spider monkey, the tip of the tail has either a bare patch or adhesive pad, which provides increased friction. Claws can be used to interact with rough substrates and reorient the direction of forces the animal applies. This is what allows squirrels to climb tree trunks that are so large to be essentially flat from the perspective of such a small animal. However, claws can interfere with an animal's ability to grasp very small branches, as they may wrap too far around and prick the animal's own paw. Frictional gripping is used by primates, relying upon hairless fingertips. Squeezing the branch between the fingertips generates frictional force that holds the animal's hand to the branch. However, this type of grip depends upon the angle of the frictional force, thus upon the diameter of the branch, with larger branches resulting in reduced gripping ability. To control descent, especially down large diameter branches, some arboreal animals such as squirrels have evolved highly mobile ankle joints that permit rotating the foot into a 'reversed' posture. This allows the claws to hook into the rough surface of the bark, opposing the force of gravity. Small size provides many advantages to arboreal species: such as increasing the relative size of branches to the animal, lower center of mass, increased stability, lower mass (allowing movement on smaller branches) and the ability to move through more cluttered habitat. Size relating to weight affects gliding animals such as the sugar glider. Some species of primate, bat and all species of sloth achieve passive stability by hanging beneath the branch. Both pitching and tipping become irrelevant, as the only method of failure would be losing their grip.

Aerial

Bats are the only mammals that can truly fly. They fly through the air at a constant speed by moving their wings up and down (usually with some fore-aft movement as well). Because the animal is in motion, there is some airflow relative to its body which, combined with the velocity of the wings, generates a faster airflow moving over the wing. This generates a lift force vector pointing forwards and upwards, and a drag force vector pointing rearwards and upwards. The upwards components of these counteract gravity, keeping the body in the air, while the forward component provides thrust to counteract both the drag from the wing and from the body as a whole.

The wings of bats are much thinner and consist of more bones than that of birds, allowing bats to maneuver more accurately and fly with more lift and less drag. By folding the wings inwards towards their body on the upstroke, they use 35% less energy during flight than birds. The membranes are delicate, ripping easily; however, the tissue of the bat's membrane is able to regrow, such that small tears can heal quickly. The surface of their wings is equipped with touch-sensitive receptors on small bumps called Merkel cells, also found on human fingertips. These sensitive areas are different in bats, as each bump has a tiny hair in the center, making it even more sensitive and allowing the bat to detect and collect information about the air flowing over its wings, and to fly more efficiently by changing the shape of its wings in response.

Fossorial

Fossorial creatures live in subterranean environments. Many fossorial mammals were classified under the, now obsolete, order Insectivora, such as shrews, hedgehogs and moles. Fossorial mammals have a fusiform body, thickest at the shoulders and tapering off at the tail and nose. Unable to see in the dark burrows, most have degenerated eyes, but degeneration varies between species; pocket gophers, for example, are only semi-fossorial and have very small yet functional eyes, in the fully fossorial marsupial mole the eyes are degenerated and useless, talpa moles have vestigial eyes and the cape golden mole has a layer of skin covering the eyes. External ears flaps are also very small or absent. Truly fossorial mammals have short, stout legs as strength is more important than speed to a burrowing mammal, but semi-fossorial mammals have cursorial legs. The front paws are broad and have strong claws to help in loosening dirt while excavating burrows, and the back paws have webbing, as well as claws, which aids in throwing loosened dirt backwards. Most have large incisors to prevent dirt from flying into their mouth.

Semi-fossorial wombat (left) vs. fully fossorial golden mole (right)

Aquatic

Fully aquatic mammals, the cetaceans and sirenians, have lost their legs and have a tail fin to propel themselves through the water. Flipper movement is continuous. Whales swim by moving their tail fin and lower body up and down, propelling themselves through vertical movement, while their flippers are mainly used for steering. Their skeletal anatomy allows them to be fast swimmers. Most species have a dorsal fin to prevent themselves from turning upside-down in the water. The flukes of sirenians are raised up and down in long strokes to move the animal forward, and can be twisted to turn. The forelimbs are paddle-like flippers which aid in turning and slowing.

Semi-aquatic mammals, like pinnipeds, have two pairs of flippers on the front and back, the fore-flippers and hind-flippers. The elbows and ankles are enclosed within the body. Pinnipeds have several adaptions for reducing drag. In addition to their streamlined bodies, they have smooth networks of muscle bundles in their skin that may increase laminar flow and make it easier for them to slip through water. They also lack arrector pili, so their fur can be streamlined as they swim. They rely on their fore-flippers for locomotion in a wing-like manner similar to penguins and sea turtles. Fore-flipper movement is not continuous, and the animal glides between each stroke. Compared to terrestrial carnivorans, the fore-limbs are reduced in length, which gives the locomotor muscles at the shoulder and elbow joints greater mechanical advantage; the hind-flippers serve as stabilizers. Other semi-aquatic mammals include beavers, hippopotamuses, otters and platypuses. Hippos are very large semi-aquatic mammals, and their barrel-shaped bodies have graviportal skeletal structures, adapted to carrying their enormous weight, and their specific gravity allows them to sink and move along the bottom of a river.

Reptiles

Reptiles are tetrapod animals in the class Reptilia, comprising today's turtles, crocodilians, snakes, amphisbaenians, lizards, tuatara, and their extinct relatives. The study of these traditional reptile orders, historically combined with that of modern amphibians, is called herpetology.

Because some reptiles are more closely related to birds than they are to other reptiles (e.g., crocodiles are more closely related to birds than they are to lizards), the traditional groups of "reptiles" listed above do not together constitute a monophyletic grouping or clade (consisting of all descendants of a common ancestor). For this reason, many modern scientists prefer to consider the birds part of Reptilia as well, thereby making Reptilia a monophyletic class, including all living Diapsids.

The earliest known proto-reptiles originated around 312 million years ago during the Carboniferous period, having evolved from advanced reptiliomorph tetrapods that became increasingly adapted to life on dry land. Some early examples include the lizard-like *Hylonomus* and *Casineria*. In addition to the living reptiles, there are many diverse groups that are now extinct, in some cases due to mass extinction events. In particular, the Cretaceous–Paleogene extinction event wiped out the pterosaurs, plesiosaurs, ornithischians, and sauropods, as well as many species of theropods, including troodontids, dromaeosaurids, tyrannosaurids, and abelisaurids, along with many Crocodyliformes, and squamates (e.g. mosasaurids).

Modern non-avian reptiles inhabit all the continents except Antarctica, although some birds are found on the periphery of Antarctica. Several living subgroups are recognized: Testudines (turtles and tortoises), 350 species; Rhynchocephalia (tuatara from New Zealand), 1 species; Squamata (lizards, snakes, and worm lizards), over 10,200 species; Crocodilia (crocodiles, gavials, caimans, and alligators), 24 species; and Aves (birds), approximately 10,000 species.

Reptiles are tetrapod vertebrates, creatures that either have four limbs or, like snakes, are descended from four-limbed ancestors. Unlike amphibians, reptiles do not have an aquatic larval stage. Most reptiles are oviparous, although several species of squamates are viviparous, as were some extinct aquatic clades—the fetus develops within the mother, contained in a placenta rather than an eggshell. As amniotes, reptile eggs are surrounded by membranes for protection and transport, which adapt them to reproduction on dry land. Many of the viviparous species feed their fetuses through various forms of placenta analogous to those of mammals, with some providing initial care for their hatchlings. Extant reptiles range in size from a tiny gecko, *Sphaerodactylus ariasae*, which can grow up to 17 mm (0.7 in) to the saltwater crocodile, *Crocodylus porosus*, which may reach 6 m (19.7 ft) in length and weigh over 1,000 kg (2,200 lb).

Morphology and Physiology

Circulation

Thermographic image of monitor lizards

All squamates and turtles have a three-chambered heart consisting of two atria, one variably partitioned ventricle, and two aortas that lead to the systemic circulation. The degree of mixing of oxygenated and deoxygenated blood in the three-chambered heart varies depending on the species and physiological state. Under different conditions, deoxygenated blood can be shunted back to the body or oxygenated blood can be shunted back to the lungs. This variation in blood flow has been hypothesized to allow more effective thermoregulation and longer diving times for aquatic species, but has not been shown to be a fitness advantage.

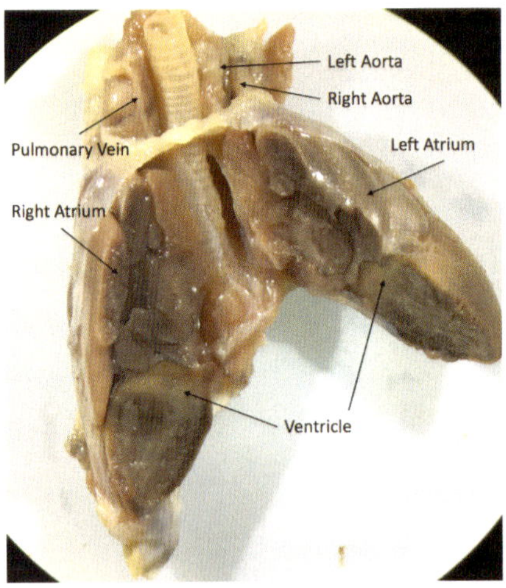

A juvenile Iguana heart bisected through the ventricle, bisecting the left and right atrium.

For example, Iguana hearts, like the majority of the squamates hearts, are composed of three chambers with two aorta and one ventricle, cardiac involuntary muscles. The main structures of the heart are the sinus venosus, the pacemaker, the left atrium, the right atruim, the atrioventricular valve, the cavum venosum, cavum arteriosum, the cavum pulmonale, the muscular ridge, the ventricular ridge, pulmonary veins, and paired aortic arches.

Some squamate species (e.g., pythons and monitor lizards) have three-chambered hearts that become functionally four-chambered hearts during contraction. This is made possible by a muscular ridge that subdivides the ventricle during ventricular diastole and completely divides it during ventricular systole. Because of this ridge, some of these squamates are capable of producing ventricular pressure differentials that are equivalent to those seen in mammalian and avian hearts.

Crocodilians have an anatomically four-chambered heart, similar to birds, but also have two systemic aortas and are therefore capable of bypassing their pulmonary circulation.

Metabolism

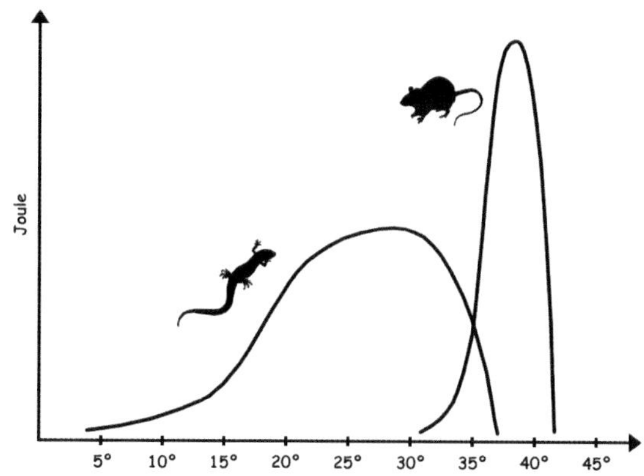

Sustained energy output (joules) of a typical reptile versus a similar size mammal as a function of core body temperature. The mammal has a much higher peak output, but can only function over a very narrow range of body temperature.

Modern non-avian reptiles exhibit some form of cold-bloodedness (i.e. some mix of poikilothermy, ectothermy, and bradymetabolism) so that they have limited physiological means of keeping the body temperature constant and often rely on external sources of heat. Due to a less stable core temperature than birds and mammals, reptilian biochemistry requires enzymes capable of maintaining efficiency over a greater range of temperatures than in the case for warm-blooded animals. The optimum body temperature range varies with species, but is typically below that of warm-blooded animals; for many lizards, it falls in the 24°–35 °C (75°–95 °F) range, while extreme heat-adapted species, like the American desert iguana *Dipsosaurus dorsalis*, can have optimal physiological temperatures in the mammalian range, between 35° and 40 °C (95° and 104 °F). While the optimum temperature is often encountered when the animal is active, the low basal metabolism makes body temperature drop rapidly when the animal is inactive.

As in all animals, reptilian muscle action produces heat. In large reptiles, like leatherback turtles, the low surface-to-volume ratio allows this metabolically produced heat to keep the animals warmer than their environment even though they do not have a warm-blooded metabolism. This form of homeothermy is called gigantothermy; it has been suggested as having been common in large dinosaurs and other extinct large-bodied reptiles.

The benefit of a low resting metabolism is that it requires far less fuel to sustain bodily functions. By using temperature variations in their surroundings, or by remaining cold when they do not need to move, reptiles can save considerable amounts of energy compared to endothermic animals of the same size. A crocodile needs from a tenth to a fifth of the food necessary for a lion of the same weight and can live half a year without eating. Lower food requirements and adaptive metabolisms allow reptiles to dominate the animal life in regions where net calorie availability is too low to sustain large-bodied mammals and birds.

It is generally assumed that reptiles are unable to produce the sustained high energy output necessary for long distance chases or flying. Higher energetic capacity might have been responsible for the evolution of warm-bloodedness in birds and mammals. However, investigation of correlations between active capacity and thermophysiology show a weak relationship. Most extant reptiles are carnivores with a sit-and-wait feeding strategy; whether reptiles are cold blooded due to their ecology is not clear. Energetic studies on some reptiles have shown active capacities equal to or greater than similar sized warm-blooded animals.

Respiratory System

All reptiles breathe using lungs. Aquatic turtles have developed more permeable skin, and some species have modified their cloaca to increase the area for gas exchange. Even with these adaptations, breathing is never fully accomplished without lungs. Lung ventilation is accomplished differently in each main reptile group. In squamates, the lungs are ventilated almost exclusively by the axial musculature. This is also the same musculature that is used during locomotion. Because of this constraint, most squamates are forced to hold their breath during intense runs. Some, however, have found a way around it. Varanids, and a few other lizard species, employ buccal pumping as a complement to their normal "axial breathing". This allows the animals to completely fill their lungs during intense locomotion, and thus remain aerobically active for a long time. Tegu lizards are known to possess a proto-diaphragm, which separates the pulmonary cavity from the visceral cavity. While not actually capable of movement, it does allow for greater lung inflation, by taking the weight of the viscera off the lungs.

Crocodilians actually have a muscular diaphragm that is analogous to the mammalian diaphragm. The difference is that the muscles for the crocodilian diaphragm pull the pubis (part of the pelvis, which is movable in crocodilians) back, which brings the liver down, thus freeing space for the lungs to expand. This type of diaphragmatic setup has been referred to as the "hepatic piston". The airways form a number of double tubular chambers within each lung. On inhalation and exhalation air moves through the airways in the same direction, thus creating a unidirectional airflow through the lungs. A similar system is found in birds, monitor lizards and iguanas.

Most reptiles lack a secondary palate, meaning that they must hold their breath while swallowing. Crocodilians have evolved a bony secondary palate that allows them to continue breathing while remaining submerged (and protect their brains against damage by struggling prey). Skinks (family Scincidae) also have evolved a bony secondary palate, to varying degrees. Snakes took a different approach and extended their trachea instead. Their tracheal extension sticks out like a fleshy straw, and allows these animals to swallow large prey without suffering from asphyxiation.

Turtles and Tortoises

Red-eared slider taking a gulp of air

How turtles and tortoises breathe has been the subject of much study. To date, only a few species have been studied thoroughly enough to get an idea of how those turtles breathe. The varied results indicate that turtles and tortoises have found a variety of solutions to this problem.

The difficulty is that most turtle shells are rigid and do not allow for the type of expansion and contraction that other amniotes use to ventilate their lungs. Some turtles, such as the Indian flapshell (*Lissemys punctata*), have a sheet of muscle that envelops the lungs. When it contracts, the turtle can exhale. When at rest, the turtle can retract the limbs into the body cavity and force air out of the lungs. When the turtle protracts its limbs, the pressure inside the lungs is reduced, and the turtle can suck air in. Turtle lungs are attached to the inside of the top of the shell (carapace), with the bottom of the lungs attached (via connective tissue) to the rest of the viscera. By using a series of special muscles (roughly equivalent to a diaphragm), turtles are capable of pushing their viscera up and down, resulting in effective respiration, since many of these muscles have attachment points in conjunction with their forelimbs (indeed, many of the muscles expand into the limb pockets during contraction).

Breathing during locomotion has been studied in three species, and they show different patterns. Adult female green sea turtles do not breathe as they crutch along their nesting beaches. They hold their breath during terrestrial locomotion and breathe in bouts as they rest. North American box turtles breathe continuously during locomotion, and the ventilation cycle is not coordinated with the limb movements. This is because they use their abdominal muscles to breathe during locomotion. The last species to have been studied is the red-eared slider, which also breathes during locomotion, but takes smaller breaths during locomotion than during small pauses between locomotor bouts, indicating that there may be mechanical interference between the limb movements and the breathing apparatus. Box turtles have also been observed to breathe while completely sealed up inside their shells.

Skin

Skin of a sand lizard, showing squamate reptiles iconic scales

Reptilian skin is covered in a horny epidermis, making it watertight and enabling reptiles to live on dry land, in contrast to amphibians. Compared to mammalian skin, that of reptiles is rather thin and lacks the thick dermal layer that produces leather in mammals. Exposed parts of reptiles are protected by scales or scutes, sometimes with a bony base, forming armor. In lepidosaurians, such as lizards and snakes, the whole skin is covered in overlapping epidermal scales. Such scales were once thought to be typical of the class Reptilia as a whole, but are now known to occur only in lepidosaurians. The scales found in turtles and crocodiles are of dermal, rather than epidermal, origin and are properly termed scutes. In turtles, the body is hidden inside a hard shell composed of fused scutes.

Lacking a thick dermis, reptilian leather is not as strong as mammalian leather. It is used in leather-wares for decorative purposes for shoes, belts and handbags, particularly crocodile skin.

Shedding. Reptiles shed their skin through a process called ecdysis which occurs continuously throughout their lifetime. In particular, younger reptiles tend to shed once every 5–6 weeks while adults shed 3-4 times a year. Younger reptiles shed more because of their rapid growth rate. Once full size, the frequency of shedding drastically decreases. The process of ecdysis involves forming a new layer of skin under the old one. Proteolytic enzymes and lymphatic fluid is secreted between the old and new layers of skin. Consequently, this lifts the old skin from the new one allowing shedding to occur. Snakes will shed from the head to the tail while lizards shed in a "patchy pattern". Dysecdysis, a common skin disease in snakes and lizards, will occur when ecdysis, or shedding, fails. There are numerous reasons why shedding fails and can be related to inadequate humidity and temperature, nutritional deficiencies, dehydration and traumatic injuries. Nutritional deficiencies decrease proteolytic enzymes while dehydration reduces lymphatic fluids to separate the skin layers. Traumatic injuries on the other hand, form scars that will not allow new scales to form and disrupt the process of ecdysis.

Excretion

Excretion is performed mainly by two small kidneys. In diapsids, uric acid is the main nitrogenous waste product; turtles, like mammals, excrete mainly urea. Unlike the kidneys of mammals and birds, reptile kidneys are unable to produce liquid urine more concentrated than their body fluid. This is because they lack a specialized structure called a loop of Henle, which is present in the nephrons of birds and mammals. Because of this, many reptiles use the colon to aid in the reabsorption of water. Some are also able to take up water stored in the bladder. Excess salts are also excreted by nasal and lingual salt glands in some reptiles.

In all reptiles the urinogenital ducts and the anus both empty into an organ called a cloaca. In some reptiles, a midventral wall in the cloaca may open into a urinary bladder, but not all. It is present in all turtles and tortoises as well as most lizards, but is lacking in the monitor lizard, the legless lizards. It is absent in the snakes, alligators, and crocodiles.

Many turtles, tortoises, and lizards have proportionally very large bladders. Charles Darwin noted that the Galapagos tortoise had a bladder which could store up to 20% of its body weight. Such adaptations are the result of environments such as remote islands and deserts where water is very scarce. Other desert-dwelling reptiles have large bladders that can store a long-term reservoir of water for up to several months and aid in osmoregulation.

Turtles have two or more accessory urinary bladders, located lateral to the neck of the urinary bladder and dorsal to the pubis, occupying a significant portion of their body cavity. Their bladder is also usually bilobed with a left and right section. The right section is located under the liver, which prevents large stones from remaining in that side while the left section is more likely to have calculi.

Digestion

A colubrid snake, Dolichophis jugularis, eating a legless lizard, Pseudopus apodus. Most reptiles are carnivorous, and many primarily eat other reptiles.

Gastroliths from a plesiosaur

Most reptiles are insectivorous or carnivorous and have simple and comparatively short digestive tracts due to meat being fairly simple to break down and digest. Digestion is slower than in mammals, reflecting their lower resting metabolism and their inability to divide and masticate their food. Their poikilotherm metabolism has very low energy requirements, allowing large reptiles like crocodiles and large constrictors to live from a single large meal for months, digesting it slowly.

While modern reptiles are predominantly carnivorous, during the early history of reptiles several groups produced some herbivorous megafauna: in the Paleozoic, the pareiasaurs; and in the Mesozoic several lines of dinosaurs. Today, turtles are the only predominantly herbivorous reptile group, but several lines of agamas and iguanas have evolved to live wholly or partly on plants.

Herbivorous reptiles face the same problems of mastication as herbivorous mammals but, lacking the complex teeth of mammals, many species swallow rocks and pebbles (so called gastroliths) to aid in digestion: The rocks are washed around in the stomach, helping to grind up plant matter. Fossil gastroliths have been found associated with both ornithopods and sauropods, though whether they actually functioned as a gastric mill in the latter is disputed. Salt water crocodiles also use gastroliths as ballast, stabilizing them in the water or helping them to dive. A dual function as both stabilizing ballast and digestion aid has been suggested for gastroliths found in plesiosaurs.

Nerves

The reptilian nervous system contains the same basic part of the amphibian brain, but the reptile cerebrum and cerebellum are slightly larger. Most typical sense organs are well developed with certain exceptions, most notably the snake's lack of external ears (middle and inner ears are present). There are twelve pairs of cranial nerves. Due to their short cochlea, reptiles use electrical tuning to expand their range of audible frequencies.

Intelligence

Reptiles are generally considered less intelligent than mammals and birds. The size of their brain relative to their body is much less than that of mammals, the encephalization quotient being about one tenth of that of mammals, though larger reptiles can show more complex brain development. Larger lizards, like the monitors, are known to exhibit complex behavior, including cooperation. Crocodiles have relatively larger brains and show a fairly complex social structure. The Komodo dragon is even known to engage in play, as are turtles, which are also considered to be social creatures, and sometimes switch between monogamy and promiscuity in their sexual behavior. One study found that wood turtles were better than white rats at learning to navigate mazes.

Vision

Most reptiles are diurnal animals. The vision is typically adapted to daylight conditions, with color vision and more advanced visual depth perception than in amphibians and most mammals. In some species, such as blind snakes, vision is reduced.

Some snakes have extra sets of visual organs in the form of pits sensitive to infrared radiation (heat). Such heat-sensitive pits are particularly well developed in the pit vipers, but are also found in boas and pythons. These pits allow the snakes to sense the body heat of birds and mammals, enabling pit vipers to hunt rodents in the dark.

Reproduction

Reptiles generally reproduce sexually, though some are capable of asexual reproduction. All reproductive activity occurs through the cloaca, the single exit/entrance at the base of the tail where waste is also eliminated. Most reptiles have copulatory organs, which are usually retracted or inverted and stored inside the body. In turtles and crocodilians, the male has a single median penis, while squamates, including snakes and lizards, possess a pair of hemipenes, only one of which is typically used in each session. Tuatara, however, lack copulatory organs, and so the male and female simply press their cloacas together as the male discharges sperm.

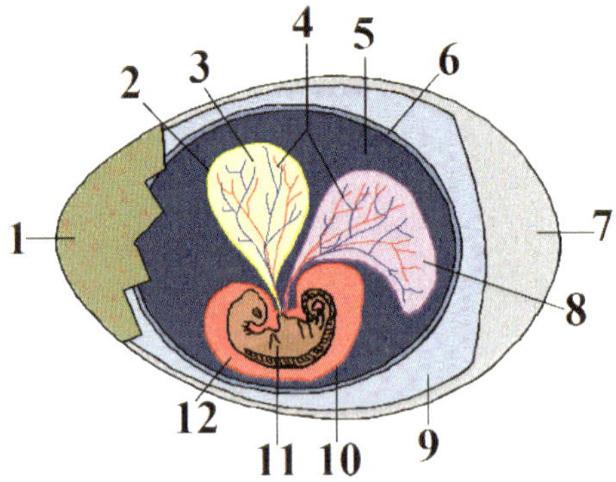

Crocodilian egg diagram
1. eggshell, 2. yolk sac, 3. yolk (nutrients), 4. vessels, 5. amnion, 6. chorion, 7. air space, 8. allantois, 9. albumin (egg white), 10. amniotic sac, 11. crocodile embryo, 12. amniotic fluid

Most reptiles reproduce sexually, for example this *Trachylepis maculilabris* skink

Geckos mating

Most reptiles lay amniotic eggs covered with leathery or calcareous shells. An amnion, chorion, and allantois are present during embryonic life. The eggshell (1) protects the crocodile embryo (11) and keeps it from drying out, but it is flexible to allow gas exchange. The chorion (6) aids in gas exchange between the inside and outside of the egg. It allows carbon dioxide to exit the egg and oxygen gas to enter the egg. The albumin (9) further protects the embryo and serves as a reservoir for water and protein.

The allantois (8) is a sac that collects the metabolic waste produced by the embryo. The amniotic sac (10) contains amniotic fluid (12) which protects and cushions the embryo. The amnion (5) aids in osmoregulation and serves as a saltwater reservoir. The yolk sac (2) surrounding the yolk (3) contains protein and fat rich nutrients that are absorbed by the embryo via vessels (4) that allow the embryo to grow and metabolize. The air space (7) provides the embryo with oxygen while it is hatching. This ensures that the embryo will not suffocate while it is hatching. There are no larval stages of development. Viviparity and ovoviviparity have evolved in many extinct clades of reptiles and in squamates. In the latter group, many species, including all boas and most vipers, utilize this mode of reproduction. The degree of viviparity varies; some species simply retain the eggs until just before hatching, others provide maternal nourishment to supplement the yolk, and yet others lack any yolk and provide all nutrients via a structure similar to the mammalian placenta. The earliest documented case of viviparity in reptiles is the Early Permian mesosaurs, although some individuals or taxa in that clade may also have been oviparous because a putative isolated egg has also been found. Several groups of Mesozoic marine reptiles also exhibited viviparity, such as mosasaurs, ichthyosaurs, and Sauropterygia, a group that include pachypleurosaurs and Plesiosauria.

Reptiles have amniotic eggs with hard or leathery shells, requiring internal fertilization when mating.

Asexual reproduction has been identified in squamates in six families of lizards and one snake. In some species of squamates, a population of females is able to produce a unisexual diploid clone of the mother. This form of asexual reproduction, called parthenogenesis, occurs in several species of gecko, and is particularly widespread in the teiids (especially *Aspidocelis*) and lacertids (*Lacerta*). In captivity, Komodo dragons (Varanidae) have reproduced by parthenogenesis.

Parthenogenetic species are suspected to occur among chameleons, agamids, xantusiids, and typhlopids.

Some reptiles exhibit temperature-dependent sex determination (TDSD), in which the incubation temperature determines whether a particular egg hatches as male or female. TDSD is most common in turtles and crocodiles, but also occurs in lizards and tuatara. To date, there has been no confirmation of whether TDSD occurs in snakes.

Birds

Bird, (class Aves), any of the more than 10,400 living species unique in having feathers, the major characteristic that distinguishes them from all other animals. A more-elaborate definition would note that they are warm-blooded vertebrates more related to reptiles than to mammals and that they have a four-chambered heart (as do mammals), forelimbs modified into wings (a trait shared with bats), a hard-shelled egg, and keen vision, the major sense they rely on for information about the environment. Their sense of smell is not highly developed, and auditory range is limited. Most birds are diurnal in habit. More than 1,000 extinct species have been identified from fossil remains.

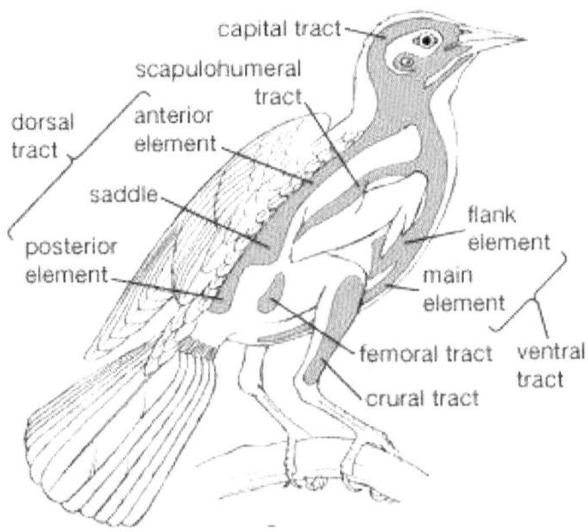

Basic body feather tracts on a generalized songbird. The shaded areas show the right half of each tract.

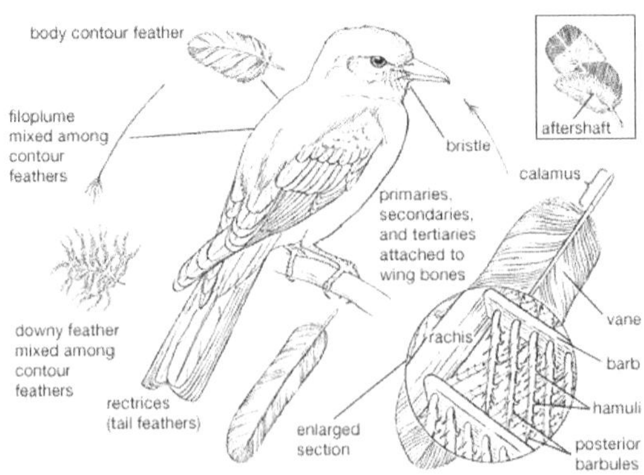

Feather types and their distribution on a typical perching bird.

Since earliest times birds have been not only a material but also a cultural resource. Bird figures were created by prehistoric humans in the Lascaux Grotto of France and have featured prominently in the mythology and literature of societies throughout the world. Long before ornithology was practiced as a science, interest in birds and the knowledge of them found expression in conversation and stories, which then crystallized into the records of general culture. Ancient Egyptian hieroglyphs and paintings, for example, include bird figures. The Bible refers to Noah's use of the raven and dove to bring him information about the proverbial Flood.

Various bird attributes, real or imagined, have led to their symbolic use in language as in art. Aesop's fables abound in bird characters. The *Physiologus* and its descendants, the bestiaries of the Middle Ages, contain moralistic writings that use birds as symbols for conveying ideas but indicate little knowledge of the birds themselves. Supernatural beliefs about birds probably took hold as early as recognition of the fact that some birds were good to eat. Australian Aborigines, for example, drove the black-and-white flycatcher from camp, lest it overhear conversation and carry the tales to enemies. Peoples of the Pacific Islands saw frigate birds as symbols of the Sun and as carriers of omens and frequently portrayed them in their art. The raven—a common symbol of dark prophecy—was the most important creature to the Indians of the Pacific Northwest and was immortalized in Edgar Allan Poe's poem "The Raven." Eagles have long been symbols of power and prestige in many parts of the world, including Europe, where their representations are often seen in heraldry. Native Americans sprinkled eagle down before guests as a sign of peace and friendship, and eagle feathers were commonly used in rituals and headdresses. The resplendent quetzal—the national bird of Guatemala, which shares its name with the currency and is a popular motif in art, fabric, and jewelry—was worshipped and deified by the ancient Mayans and Aztecs. Highly symbolic birds include the phoenix, representing resurrection, and the owl, a common symbol of wisdom but also a reminder of death in Native American mythology. The bird in general has long been a common Christian symbol of the transcendent soul, and in medieval iconography a bird entangled in foliage symbolized the soul embroiled in the materialism of the secular world.

In modern times the recreational pleasures of bird-watching have grown in tandem with the rise of environmentalism. Evolving from the American and European "shoot-and-stuff" mania of the 19th century, bird-watching became a sportlike activity based on rapid identification—the rarest being the most rewarding—with the aid of binoculars and spotting scopes. The change from shooting to sighting coincided with campaigns, beginning about 1900, to halt the slaughter of wild birds for food and millinery. Bird-watching was advanced by the publication of excellent field guides and improvements in photography and sound recording. By mid-century the watcher's enjoyable but rather unsophisticated tallying of "year lists" and "life lists" of species personally observed was being augmented, if not replaced, by interest in careful studies of bird behaviour, migration, ecology, and the like. This trend was abetted by bird banding (called ringing in the United Kingdom) and by such organizations as the British Trust for Ornithology and the National Audubon Society, which coordinate professional and amateur observations and efforts with scientific studies.

Anatomy and Physiology

Compared with other vertebrates, birds have a body plan that shows many unusual adaptations, mostly to facilitate flight.

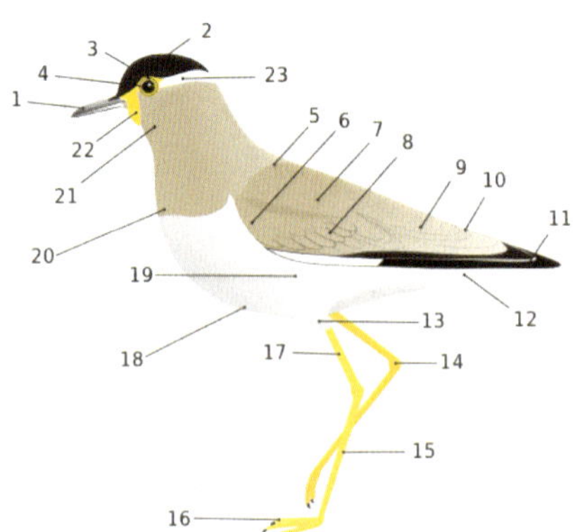

External anatomy of a bird (example: yellow-wattled lapwing).

1 Beak, 2 Head, 3 Iris, 4 Pupil, 5 Mantle, 6 Lesser coverts, 7 Scapulars, 8 Median coverts, 9 Tertials, 10 Rump, 11 Primaries, 12 Vent, 13 Thigh, 14 Tibio-tarsal articulation, 15 Tarsus, 16 Foot, 17 Tibia, 18 Belly, 19 Flanks, 20 Breast, 21 Throat, 22 Wattle, 23 Eyestripe

Skeletal System

The skeleton consists of very lightweight bones. They have large air-filled cavities (called pneumatic cavities) which connect with the respiratory system. The skull bones in adults are fused and do not show cranial sutures. The orbits are large and separated by a bony septum. The spine has cervical, thoracic, lumbar and caudal regions with the number of cervical (neck) vertebrae highly variable and especially flexible, but movement is reduced in the anterior thoracic vertebrae and absent in the later vertebrae. The last few are fused with the pelvis to form the synsacrum. The ribs are flattened and the sternum is keeled for the attachment of flight muscles except in the flightless bird orders. The forelimbs are modified into wings.

Excretory System

Like the reptiles, birds are primarily uricotelic, that is, their kidneys extract nitrogenous waste from their bloodstream and excrete it as uric acid instead of urea or ammonia through the ureters into the intestine. Birds do not have a urinary bladder or external urethral opening and (with exception of the ostrich) uric acid is excreted along with faeces as a semisolid waste. However, birds such as hummingbirds can be facultatively ammonotelic, excreting most of the nitrogenous wastes as ammonia. They also excrete creatine, rather than creatinine like mammals. This material, as well as the output of the intestines, emerges from the bird's cloaca. The cloaca is a multi-purpose opening: waste is expelled through it, most birds mate by joining cloaca, and females lay eggs from it. In addition, many species of birds regurgitate pellets.

Males within Palaeognathae (with the exception of the kiwis), the Anseriformes (with the exception of screamers), and in rudimentary forms in Galliformes (but fully developed in Cracidae) possess a penis, which is never present in Neoaves. The length is thought to be related to sperm competition. When not copulating, it is hidden within the proctodeum compartment within the cloaca, just inside the vent. The digestive system of birds is unique, with a crop for storage and a gizzard that contains swallowed stones for grinding food to compensate for the lack of teeth. Most birds are highly adapted for rapid digestion to aid with flight. Some migratory birds have adapted to use protein from many parts of their bodies, including protein from the intestines, as additional energy during migration.

Respiratory and Circulatory Systems

Birds have one of the most complex respiratory systems of all animal groups. Upon inhalation, 75% of the fresh air bypasses the lungs and flows directly into a posterior air sac which extends from the lungs and connects with air spaces in the bones and fills them with air. The other 25% of the air goes directly into the lungs. When the bird exhales, the used air flows out of the lungs and the stored fresh air from the posterior air sac is simultaneously forced into the lungs. Thus, a bird's lungs receive a constant supply of fresh air during both inhalation and exhalation. Sound production is achieved using the syrinx, a muscular chamber incorporating multiple tympanic membranes which diverges from the lower end of the trachea; the trachea being elongated in some species, increasing the volume of vocalisations and the perception of the bird's size.

In birds, the main arteries taking blood away from the heart originate from the right aortic arch (or pharyngeal arch), unlike in the mammals where the left aortic arch forms this part of the aorta. The postcava receives blood from the limbs via the renal portal system. Unlike in mammals, the circulating red blood cells in birds retain their nucleus.

Heart type and Features

Didactic model of an avian heart.

The avian circulatory system is driven by a four-chambered, myogenic heart contained in a fibrous pericardial sac. This pericardial sac is filled with a serous fluid for lubrication. The heart itself is divided into a right and left half, each with an atrium and ventricle. The atrium and ventricles of each side are separated by atrioventricular valves which prevent back flow from one chamber to the next during contraction. Being myogenic, the heart's pace is maintained by pacemaker cells found in the sinoatrial node, located on the right atrium.

The sinoatrial node uses calcium to cause a depolarising signal transduction pathway from the atrium through right and left atrioventricular bundle which communicates contraction to the ventricles. The avian heart also consists of muscular arches that are made up of thick bundles of muscular layers. Much like a mammalian heart, the avian heart is composed of endocardial, myocardial and epicardial layers. The atrium walls tend to be thinner than the ventricle walls, due to the intense ventricular contraction used to pump oxygenated blood throughout the body. Avian hearts are generally larger than mammalian hearts when compared to body mass. This adaptation allows more blood to be pumped to meet the high metabolic need associated with flight.

Organisation

Birds have a very efficient system for diffusing oxygen into the blood; birds have a ten times greater surface area to gas exchange volume than mammals. As a result, birds have more blood in their capillaries per unit of volume of lung than a mammal. The arteries are composed of thick elastic muscles to withstand the pressure of the ventricular constriction, and become more rigid as they move away from the heart. Blood moves through the arteries, which undergo vasoconstriction, and into arterioles which act as a transportation system to distribute primarily oxygen as well as nutrients to all tissues of the body. As the arterioles move away from the heart and into individual organs and tissues they are further divided to increase surface area and slow blood flow. Blood travels through the arterioles and moves into the capillaries where gas exchange can occur.

Capillaries are organized into capillary beds in tissues; it is here that blood exchanges oxygen for carbon dioxide waste. In the capillary beds blood flow is slowed to allow maximum diffusion of oxygen into the tissues. Once the blood has become deoxygenated it travels through venules then veins and back to the heart. Veins, unlike arteries, are thin and rigid as they do not need to withstand extreme pressure. As blood travels through the venules to the veins a funneling occurs called vasodilation bringing blood back to the heart. Once the blood reaches the heart it moves first into the right atrium, then the right ventricle to be pumped through the lungs for further gas exchange of carbon dioxide waste for oxygen. Oxygenated blood then flows from the lungs through the left atrium to the left ventricle where it is pumped out to the body.

Nervous System

The nervous system is large relative to the bird's size. The most developed part of the brain is the one that controls the flight-related functions, while the cerebellum coordinates movement and the cerebrum controls behaviour patterns, navigation, mating and nest building. Most birds have a poor sense of smell with notable exceptions including kiwis, New World vultures and tubenoses. The avian visual system is usually highly developed. Water birds have special flexible lenses, allowing accommodation for vision in air and water. Some species also have dual fovea. Birds are tetrachromatic, possessing ultraviolet (UV) sensitive cone cells in the eye as well as green, red and blue ones. They also have have double cones, likely to mediate achromatic vision.

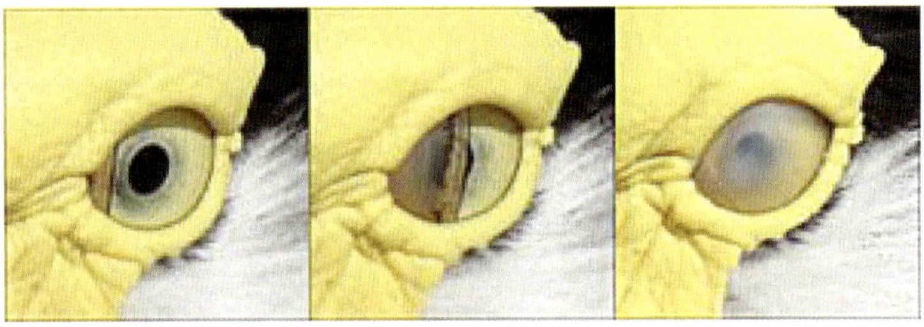

The nictitating membrane as it covers the eye of a masked lapwing

Many birds show plumage patterns in ultraviolet that are invisible to the human eye; some birds whose sexes appear similar to the naked eye are distinguished by the presence of ultraviolet reflective patches on their feathers. Male blue tits have an ultraviolet reflective crown patch which is displayed in courtship by posturing and raising of their nape feathers. Ultraviolet light is also used in foraging—kestrels have been shown to search for prey by detecting the UV reflective urine trail marks left on the ground by rodents. With the exception of pigeons and a few other species, the eyelids of birds are not used in blinking. Instead the eye is lubricated by the nictitating membrane, a third eyelid that moves horizontally. The nictitating membrane also covers the eye and acts as a contact lens in many aquatic birds. The bird retina has a fan shaped blood supply system called the pecten.

Most birds cannot move their eyes, although there are exceptions, such as the great cormorant. Birds with eyes on the sides of their heads have a wide visual field, while birds with eyes on the front of their heads, such as owls, have binocular vision and can estimate the depth of field. The avian ear lacks external pinnae but is covered by feathers, although in some birds, such as the *Asio*, *Bubo* and *Otus* owls, these feathers form tufts which resemble ears. The inner ear has a cochlea, but it is not spiral as in mammals.

Defence and Intraspecific Combat

A few species are able to use chemical defences against predators; some Procellariiformes can eject an unpleasant stomach oil against an aggressor, and some species of pitohuis from New Guinea have a powerful neurotoxin in their skin and feathers.

A lack of field observations limit our knowledge, but intraspecific conflicts are known to sometimes result in injury or death. The screamers (Anhimidae), some jacanas (*Jacana*, *Hydrophasianus*), the spur-winged goose (*Plectropterus*), the torrent duck (*Merganetta*) and nine species of lapwing (*Vanellus*) use a sharp spur on the wing as a weapon. The steamer ducks (*Tachyeres*), geese and swans (*Anserinae*), the solitaire (*Pezophaps*), sheathbills (*Chionis*), some guans (*Crax*) and stone curlews (*Burhinus*) use a bony knob on the alular metacarpal to punch and hammer opponents. The jacanas *Actophilornis* and *Irediparra* have an expanded, blade-like radius. The extinct *Xenicibis* was unique in having an elongate forelimb and massive hand which likely functioned in combat or defence as a jointed club or flail. Swans, for instance, may strike with the bony spurs and bite when defending eggs or young.

Chromosomes

Birds have two sexes: either female or male. The sex of birds is determined by the Z and W sex chromosomes, rather than by the X and Y chromosomes present in mammals. Male birds have two Z chromosomes (ZZ), and female birds have a W chromosome and a Z chromosome (WZ).

In nearly all species of birds, an individual's sex is determined at fertilisation. However, one recent study demonstrated temperature-dependent sex determination among the Australian brushturkey, for which higher temperatures during incubation resulted in a higher female-to-male sex ratio. This, however, was later proven to not be the case. These birds do not exhibit temperature-dependent sex determination, but temperature-dependent sex mortality.

Feathers, Plumage, and Scales

The disruptively patterned plumage of the African scops owl allows it to blend in with its surroundings.

Feathers are a feature characteristic of birds (though also present in some dinosaurs not currently considered to be true birds). They facilitate flight, provide insulation that aids in thermoregulation, and are used in display, camouflage, and signalling. There are several types of feathers, each serving its own set of purposes. Feathers are epidermal growths attached to the skin and arise only in specific tracts of skin called pterylae. The distribution pattern of these feather tracts (pterylosis) is used in taxonomy and systematics. The arrangement and appearance of feathers on the body, called plumage, may vary within species by age, social status, and sex.

Plumage is regularly moulted; the standard plumage of a bird that has moulted after breeding is known as the "*non-breeding*" plumage, or—in the Humphrey-Parkes terminology—"basic" plumage; breeding plumages or variations of the basic plumage are known under the Humphrey-Parkes system as "*alternate*" plumages. Moulting is annual in most species, although some may have two moults a year, and large birds of prey may moult only once every few years. Moulting patterns vary across species. In passerines, flight feathers are replaced one at a time with the innermost *primary* being the first. When the fifth of sixth primary is replaced, the outermost *tertiaries* begin to drop. After the innermost tertiaries are moulted, the *secondaries* starting from the innermost begin to drop and this proceeds to the outer feathers (centrifugal moult). The greater primary *coverts* are moulted in synchrony with the primary that they overlap.

A small number of species, such as ducks and geese, lose all of their flight feathers at once, temporarily becoming flightless. As a general rule, the tail feathers are moulted and replaced starting with the innermost pair. Centripetal moults of tail feathers are however seen in the Phasianidae. The centrifugal moult is modified in the tail feathers of woodpeckers and treecreepers, in that it begins with the second innermost pair of feathers and finishes with the central pair of feathers so that the bird maintains a functional climbing tail. The general pattern seen in passerines is that the primaries are replaced outward, secondaries inward, and the tail from centre outward. Before nesting, the females of most bird species gain a bare brood patch by losing feathers close to the belly. The skin there is well supplied with blood vessels and helps the bird in incubation.

Red lory preening

Feathers require maintenance and birds preen or groom them daily, spending an average of around 9% of their daily time on this. The bill is used to brush away foreign particles and to apply waxy secretions from the uropygial gland; these secretions protect the feathers' flexibility and act as an antimicrobial agent, inhibiting the growth of feather-degrading bacteria. This may be supplemented with the secretions of formic acid from ants, which birds receive through a behaviour known as anting, to remove feather parasites.

The scales of birds are composed of the same keratin as beaks, claws, and spurs. They are found mainly on the toes and metatarsus, but may be found further up on the ankle in some birds. Most bird scales do not overlap significantly, except in the cases of kingfishers and woodpeckers. The scales of birds are thought to be homologous to those of reptiles and mammals.

Flight

Restless flycatcher in the downstroke of flapping flight

Most birds can fly, which distinguishes them from almost all other vertebrate classes. Flight is the primary means of locomotion for most bird species and is used for searching for food and for escaping from predators. Birds have various adaptations for flight, including a lightweight skeleton, two large flight muscles, the pectoralis (which accounts for 15% of the total mass of the bird) and the supracoracoideus, as well as a modified forelimb (wing) that serves as an aerofoil.

Wing shape and size generally determine a bird's flight style and performance; many birds combine powered, flapping flight with less energy-intensive soaring flight. About 60 extant bird species are flightless, as were many extinct birds. Flightlessness often arises in birds on isolated islands, probably due to limited resources and the absence of land predators. Though flightless, penguins use similar musculature and movements to "fly" through the water, as do auks, shearwaters and dippers.

Amphibians

Amphibians are ectothermic, tetrapod vertebrates of the class Amphibia. Modern amphibians are all Lissamphibia. They inhabit a wide variety of habitats, with most species living within terrestrial, fossorial, arboreal or freshwater aquatic ecosystems. Thus amphibians typically start out as larvae living in water, but some species have developed behavioural adaptations to bypass this. The young generally undergo metamorphosis from larva with gills to an adult air-breathing form with lungs. Amphibians use their skin as a secondary respiratory surface and some small terrestrial salamanders and frogs lack lungs and rely entirely on their skin. They are superficially similar to lizards but, along with mammals and birds, reptiles are amniotes and do not require water bodies in which to breed. With their complex reproductive needs and permeable skins, amphibians are often ecological indicators; in recent decades there has been a dramatic decline in amphibian populations for many species around the globe.

The earliest amphibians evolved in the Devonian period from sarcopterygian fish with lungs and bony-limbed fins, features that were helpful in adapting to dry land. They diversified and became dominant during the Carboniferous and Permian periods, but were later displaced by reptiles and other vertebrates. Over time, amphibians shrank in size and decreased in diversity, leaving only the modern subclass Lissamphibia.

The three modern orders of amphibians are Anura (the frogs and toads), Urodela (the salamanders), and Apoda (the caecilians). The number of known amphibian species is approximately 7,000, of which nearly 90% are frogs. The smallest amphibian (and vertebrate) in the world is a frog from New Guinea (*Paedophryne amauensis*) with a length of just 7.7 mm (0.30 in). The largest living amphibian is the 1.8 m (5 ft 11 in) Chinese giant salamander (*Andrias davidianus*), but this is dwarfed by the extinct 9 m (30 ft) *Prionosuchus* from the middle Permian of Brazil. The study of amphibians is called batrachology, while the study of both reptiles and amphibians is called herpetology.

Anatomy and Physiology

Skin

The bright colours of the common reed frog (*Hyperolius viridiflavus*) are typical of a toxic species

The integumentary structure contains some typical characteristics common to terrestrial vertebrates, such as the presence of highly cornified outer layers, renewed periodically through a moulting process controlled by the pituitary and thyroid glands. Local thickenings (often called warts) are common, such as those found on toads. The outside of the skin is shed periodically mostly in one piece, in contrast to mammals and birds where it is shed in flakes. Amphibians often eat the sloughed skin. Caecilians are unique among amphibians in having mineralized dermal scales embedded in the dermis between the furrows in the skin. The similarity of these to the scales of bony fish is largely superficial. Lizards and some frogs have somewhat similar osteoderms forming bony deposits in the dermis, but this is an example of convergent evolution with similar structures having arisen independently in diverse vertebrate lineages.

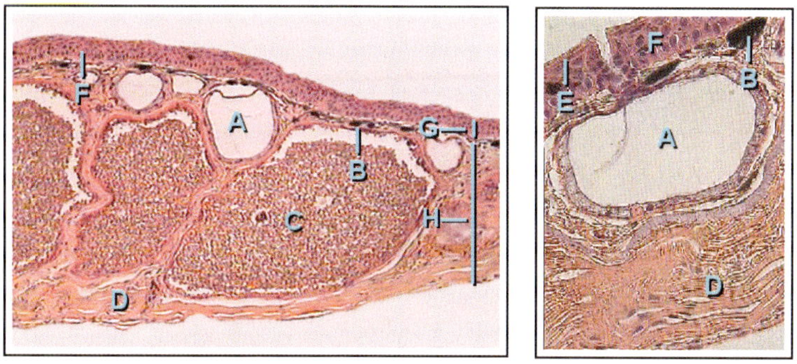

Cross section of frog skin. A: Mucus gland, B: Chromatophore, C: Granular poison gland, D: Connective tissue, E: Stratum corneum, F: Transition zone, G: Epidermis, H: Dermis

Amphibian skin is permeable to water. Gas exchange can take place through the skin (cutaneous respiration) and this allows adult amphibians to respire without rising to the surface of water and to hibernate at the bottom of ponds. To compensate for their thin and delicate skin, amphibians have evolved mucous glands, principally on their heads, backs and tails. The secretions produced by these help keep the skin moist. In addition, most species of amphibian have granular glands that secrete distasteful or poisonous substances. Some amphibian toxins can be lethal to humans while others have little effect. The main poison-producing glands, the paratoids, produce the neurotoxin bufotoxin and are located behind the ears of toads, along the backs of frogs, behind the eyes of salamanders and on the upper surface of caecilians.

The skin colour of amphibians is produced by three layers of pigment cells called chromatophores. These three cell layers consist of the melanophores (occupying the deepest layer), the guanophores (forming an intermediate layer and containing many granules, producing a blue-green colour) and the lipophores (yellow, the most superficial layer). The colour change displayed by many species is initiated by hormones secreted by the pituitary gland. Unlike bony fish, there is no direct control of the pigment cells by the nervous system, and this results in the colour change taking place more slowly than happens in fish. A vividly coloured skin usually indicates that the species is toxic and is a warning sign to predators.

Skeletal System and Locomotion

Amphibians have a skeletal system that is structurally homologous to other tetrapods, though with a number of variations. They all have four limbs except for the legless caecilians and a few species of salamander with reduced or no limbs. The bones are hollow and lightweight. The musculoskeletal system is strong to enable it to support the head and body. The bones are fully ossified and the vertebrae interlock with each other by means of overlapping processes. The pectoral girdle is supported by muscle, and the well-developed pelvic girdle is attached to the backbone by a pair of sacral ribs. The ilium slopes forward and the body is held closer to the ground than is the case in mammals.

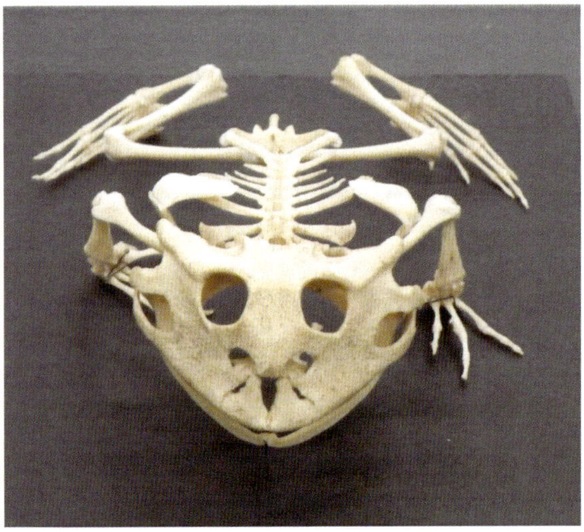

Skeleton of the Surinam horned frog (*Ceratophrys cornuta*)

In most amphibians, there are four digits on the fore foot and five on the hind foot, but no claws on either. Some salamanders have fewer digits and the amphiumas are eel-like in appearance with tiny, stubby legs. The sirens are aquatic salamanders with stumpy forelimbs and no hind limbs. The caecilians are limbless. They burrow in the manner of earthworms with zones of muscle contractions moving along the body. On the surface of the ground or in water they move by undulating their body from side to side.

In frogs, the hind legs are larger than the fore legs, especially so in those species that principally move by jumping or swimming. In the walkers and runners the hind limbs are not so large, and the burrowers mostly have short limbs and broad bodies. The feet have adaptations for the way of life, with webbing between the toes for swimming, broad adhesive toe pads for climbing, and keratinised tubercles on the hind feet for digging (frogs usually dig backwards into the soil). In most salamanders, the limbs are short and more or less the same length and project at right angles from the body. Locomotion on land is by walking and the tail often swings from side to side or is used as a prop, particularly when climbing. In their normal gait, only one leg is advanced at a time in the manner adopted by their ancestors, the lobe-finned fish. Some salamanders in the genus *Aneides* and certain plethodontids climb trees and have long limbs, large toepads and prehensile tails. In aquatic salamanders and in frog tadpoles, the tail has dorsal and ventral fins and is moved from side to side as a means of propulsion. Adult frogs do not have tails and caecilians have only very short ones.

Salamanders use their tails in defence and some are prepared to jettison them to save their lives in a process known as autotomy. Certain species in the Plethodontidae have a weak zone at the base of the tail and use this strategy readily. The tail often continues to twitch after separation which may distract the attacker and allow the salamander to escape. Both tails and limbs can be regenerated. Adult frogs are unable to regrow limbs but tadpoles can do so.

Circulatory System

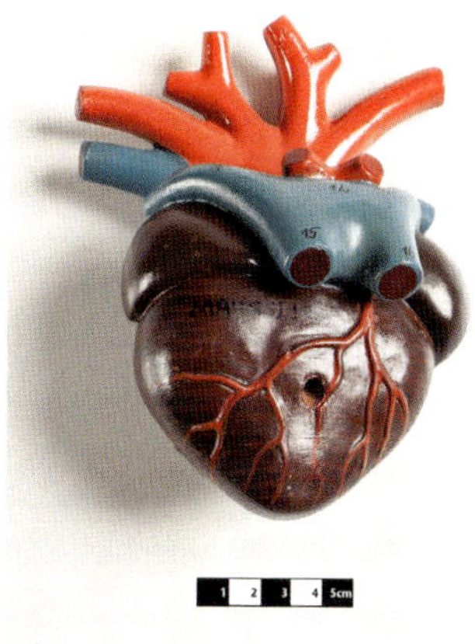

Didactic model of an amphibian heart.

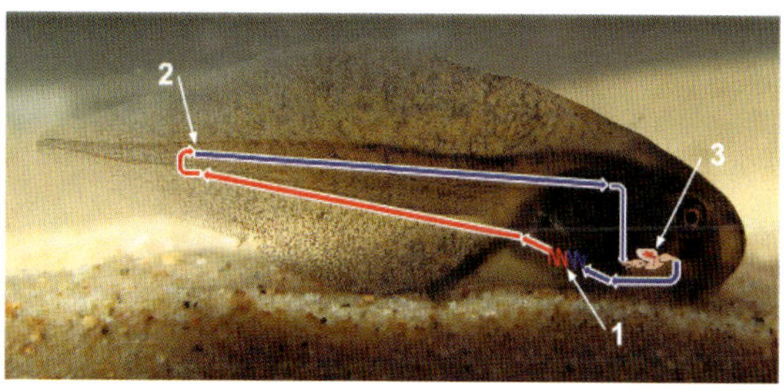

Juvenile amphibian circulatory systems are single loop systems which resemble fish. 1. Internal gills where the blood is reoxygenated 2. Point where the blood is depleted of oxygen and returns to the heart via veins 3. Two chambered heart. Red indicates oxygenated blood, and blue represents oxygen depleted blood.

Amphibians have a juvenile stage and an adult stage, and the circulatory systems of the two are distinct. In the juvenile (or tadpole) stage, the circulation is similar to that of a fish; the two-chambered heart pumps the blood through the gills where it is oxygenated, and is spread around the body and back to the heart in a single loop. In the adult stage, amphibians (especially frogs) lose their gills and develop lungs. They have a heart that consists of a single ventricle and two atria. When the ventricle starts contracting, deoxygenated blood is pumped through the pulmonary artery to the lungs. Continued contraction then pumps oxygenated blood around the rest of the body. Mixing of the two bloodstreams is minimized by the anatomy of the chambers.

Nervous and Sensory Systems

The nervous system is basically the same as in other vertebrates, with a central brain, a spinal cord, and nerves throughout the body. The amphibian brain is less well developed than that of reptiles, birds and mammals but is similar in morphology and function to that of a fish. It is believed amphibians are capable of perceiving pain. The brain consists of equal parts, cerebrum, midbrain and cerebellum. Various parts of the cerebrum process sensory input, such as smell in the olfactory lobe and sight in the optic lobe, and it is additionally the centre of behaviour and learning. The cerebellum is the center of muscular coordination and the medulla oblongata controls some organ functions including heartbeat and respiration. The brain sends signals through the spinal cord and nerves to regulate activity in the rest of the body. The pineal body, known to regulate sleep patterns in humans, is thought to produce the hormones involved in hibernation and aestivation in amphibians.

Tadpoles retain the lateral line system of their ancestral fishes, but this is lost in terrestrial adult amphibians. Some caecilians possess electroreceptors that allow them to locate objects around them when submerged in water. The ears are well developed in frogs. There is no external ear, but the large circular eardrum lies on the surface of the head just behind the eye. This vibrates and sound is transmitted through a single bone, the stapes, to the inner ear. Only high-frequency sounds like mating calls are heard in this way, but low-frequency noises can be detected through another mechanism. There is a patch of specialized haircells, called *papilla amphibiorum*, in the inner ear capable of detecting deeper sounds. Another feature, unique to frogs and salamanders, is the columella-operculum complex adjoining the auditory capsule which is involved in the transmission of both airborne and seismic signals. The ears of salamanders and caecilians are less highly developed than those of frogs as they do not normally communicate with each other through the medium of sound.

The eyes of tadpoles lack lids, but at metamorphosis, the cornea becomes more dome-shaped, the lens becomes flatter, and eyelids and associated glands and ducts develop. The adult eyes are an improvement on invertebrate eyes and were a first step in the development of more advanced vertebrate eyes. They allow colour vision and depth of focus. In the retinas are green rods, which are receptive to a wide range of wavelengths.

Digestive and Excretory Systems

Many amphibians catch their prey by flicking out an elongated tongue with a sticky tip and drawing it back into the mouth before seizing the item with their jaws. Some use inertial feeding to help them swallow the prey, repeatedly thrusting their head forward sharply causing the food to move backwards in their mouth by inertia. Most amphibians swallow their prey whole without much chewing so they possess voluminous stomachs. The short oesophagus is lined with cilia that help to move the food to the stomach and mucus produced by glands in the mouth and pharynx eases its passage. The enzyme chitinase produced in the stomach helps digest the chitinous cuticle of arthropod prey.

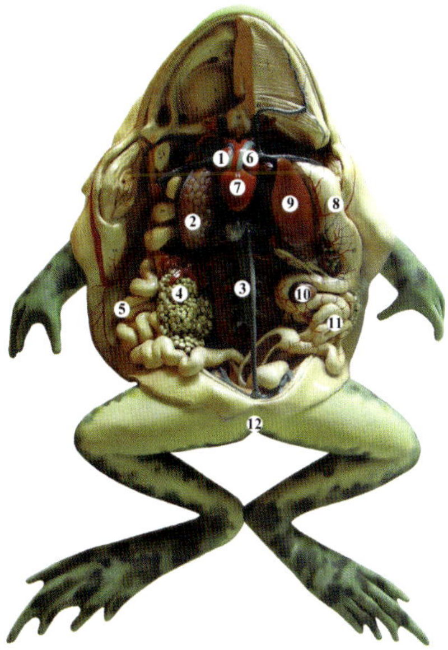

Dissected frog: 1 Right atrium, 2 Liver, 3 Aorta, 4 Egg mass, 5 Colon, 6 Left atrium, 7 Ventricle, 8 Stomach, 9 Left lung, 10 Gallbladder, 11 Small intestine, 12 Cloaca

Amphibians possess a pancreas, liver and gall bladder. The liver is usually large with two lobes. Its size is determined by its function as a glycogen and fat storage unit, and may change with the seasons as these reserves are built or used up. Adipose tissue is another important means of storing energy and this occurs in the abdomen (in internal structures called fat bodies), under the skin and, in some salamanders, in the tail.

There are two kidneys located dorsally, near the roof of the body cavity. Their job is to filter the blood of metabolic waste and transport the urine via ureters to the urinary bladder where it is stored before being passed out periodically through the cloacal vent. Larvae and most aquatic adult amphibians excrete the nitrogen as ammonia in large quantities of dilute urine, while terrestrial species, with a greater need to conserve water, excrete the less toxic product urea. Some tree frogs with limited access to water excrete most of their metabolic waste as uric acid.

Respiratory System

The lungs in amphibians are primitive compared to those of amniotes, possessing few internal septa and large alveoli, and consequently having a comparatively slow diffusion rate for oxygen entering the blood. Ventilation is accomplished by buccal pumping. Most amphibians, however, are able to exchange gases with the water or air via their skin. To enable sufficient cutaneous respiration, the surface of their highly vascularised skin must remain moist to allow the oxygen to diffuse at a sufficiently high rate. Because oxygen concentration in the water increases at both low temperatures and high flow rates, aquatic amphibians in these situations can rely primarily on cutaneous respiration, as in the Titicaca water frog and the hellbender salamander. In air, where oxygen is more concentrated, some small species can rely solely on cutaneous gas exchange, most famously the plethodontid salamanders, which have neither lungs nor gills. Many aquatic salamanders and all tadpoles have gills in their larval stage, with some (such as the axolotl) retaining gills as aquatic adults.

The axolotl (*Ambystoma mexicanum*) retains its larval form with gills into adulthood

Fishes

The first fishes are thought to have emerged some 518 million years ago during the Cambrian Period of Earth's history. Today, there more than 30,000 species of fishes found in the fresh and salt waters of the world. Living species range from the primitive, jawless lampreys and hagfishes through the cartilaginous sharks, skates, and rays to the abundant and diverse bony fishes.

Fishes range in adult length from less than 10 mm (0.4 inch) to more than 20 meters (60 feet) and in weight from about 1.5 grams (less than 0.06 ounce) to many thousands of kilograms. Some live in shallow thermal springs at temperatures slightly above 42 °C (100 °F), others in cold Arctic seas a few degrees below 0 °C (32 °F) or in cold deep waters more than 4,000 meters (13,100 feet) beneath the ocean surface.

Fish reproduction methods vary, but most fishes lay a large number of small eggs that are fertilized and scattered outside of the body. The eggs of pelagic (open ocean) fishes usually remain suspended in the open water, while many shore and freshwater fishes lay eggs on the bottom or among plants. The mortality of the young and especially of the eggs is very high, and often only a few individuals grow to maturity out of hundreds, thousands, and in some cases millions of eggs laid.

Anatomy and Physiology

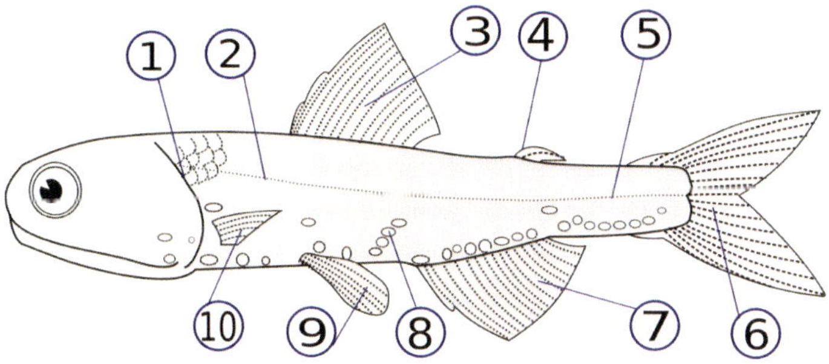

The anatomy of *Lampanyctodes hectoris* (1) – operculum (gill cover), (2) – lateral line, (3) – dorsal fin, (4) – fat fin, (5) – caudal peduncle, (6) – caudal fin, (7) – anal fin, (8) – photophores, (9) – pelvic fins (paired), (10) – pectoral fins (paired)

Respiration

Gills

Most fish exchange gases using gills on either side of the pharynx. Gills consist of threadlike structures called filaments. Each filament contains a capillary network that provides a large surface area for exchanging oxygen and carbon dioxide. Fish exchange gases by pulling oxygen-rich water through their mouths and pumping it over their gills. In some fish, capillary blood flows in the opposite direction to the water, causing countercurrent exchange. The gills push the oxygen-poor water out through openings in the sides of the pharynx. Some fish, like sharks and lampreys, possess multiple gill openings. However, bony fish have a single gill opening on each side. This opening is hidden beneath a protective bony cover called an operculum.

Juvenile bichirs have external gills, a very primitive feature that they share with larval amphibians.

Air Breathing

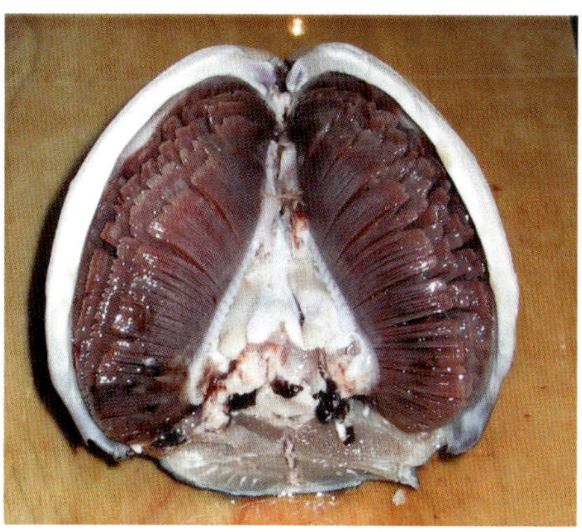

Tuna gills inside the head. The fish head is oriented snout-downwards, with the view looking towards the mouth.

Fish from multiple groups can live out of the water for extended periods. Amphibious fish such as the mudskipper can live and move about on land for up to several days, or live in stagnant or otherwise oxygen depleted water. Many such fish can breathe air via a variety of mechanisms. The skin of anguillid eels may absorb oxygen directly. The buccal cavity of the electric eel may breathe air. Catfish of the families Loricariidae, Callichthyidae, and Scoloplacidae absorb air through their digestive tracts. Lungfish, with the exception of the Australian lungfish, and bichirs have paired lungs similar to those of tetrapods and must surface to gulp fresh air through the mouth and pass spent air out through the gills. Gar and bowfin have a vascularized swim bladder that functions in the same way. Loaches, trahiras, and many catfish breathe by passing air through the gut. Mudskippers breathe by absorbing oxygen across the skin (similar to frogs). A number of fish have evolved so-called accessory breathing organs that extract oxygen from the air. Labyrinth fish (such as gouramis and bettas) have a labyrinth organ above the gills that performs this function. A few other fish have structures resembling labyrinth organs in form and function, most notably snakeheads, pikeheads, and the Clariidae catfish family.

Breathing air is primarily of use to fish that inhabit shallow, seasonally variable waters where the water's oxygen concentration may seasonally decline. Fish dependent solely on dissolved oxygen, such as perch and cichlids, quickly suffocate, while air-breathers survive for much longer, in some cases in water that is little more than wet mud. At the most extreme, some air-breathing fish are able to survive in damp burrows for weeks without water, entering a state of aestivation (summertime hibernation) until water returns.

Air breathing fish can be divided into obligate air breathers and facultative air breathers. Obligate air breathers, such as the African lungfish, *must* breathe air periodically or they suffocate. Facultative air breathers, such as the catfish *Hypostomus plecostomus*, only breathe air if they need to and will otherwise rely on their gills for oxygen. Most air breathing fish are facultative air breathers that avoid the energetic cost of rising to the surface and the fitness cost of exposure to surface predators.

Circulation

Fish have a closed-loop circulatory system. The heart pumps the blood in a single loop throughout the body. In most fish, the heart consists of four parts, including two chambers and an entrance and exit. The first part is the sinus venosus, a thin-walled sac that collects blood from the fish's veins before allowing it to flow to the second part, the atrium, which is a large muscular chamber. The atrium serves as a one-way antechamber, sends blood to the third part, ventricle. The ventricle is another thick-walled, muscular chamber and it pumps the blood, first to the fourth part, bulbus arteriosus, a large tube, and then out of the heart. The bulbus arteriosus connects to the aorta, through which blood flows to the gills for oxygenation.

Didactic model of a fish heart.

Digestion

Jaws allow fish to eat a wide variety of food, including plants and other organisms. Fish ingest food through the mouth and break it down in the esophagus. In the stomach, food is further digested and, in many fish, processed in finger-shaped pouches called pyloric caeca, which secrete digestive enzymes and absorb nutrients. Organs such as the liver and pancreas add enzymes and various chemicals as the food moves through the digestive tract. The intestine completes the process of digestion and nutrient absorption.

Excretion

As with many aquatic animals, most fish release their nitrogenous wastes as ammonia. Some of the wastes diffuse through the gills. Blood wastes are filtered by the kidneys.

Saltwater fish tend to lose water because of osmosis. Their kidneys return water to the body. The reverse happens in freshwater fish: they tend to gain water osmotically. Their kidneys produce dilute urine for excretion. Some fish have specially adapted kidneys that vary in function, allowing them to move from freshwater to saltwater.

Scales

The scales of fish originate from the mesoderm (skin); they may be similar in structure to teeth.

Sensory and Nervous System

Central Nervous System

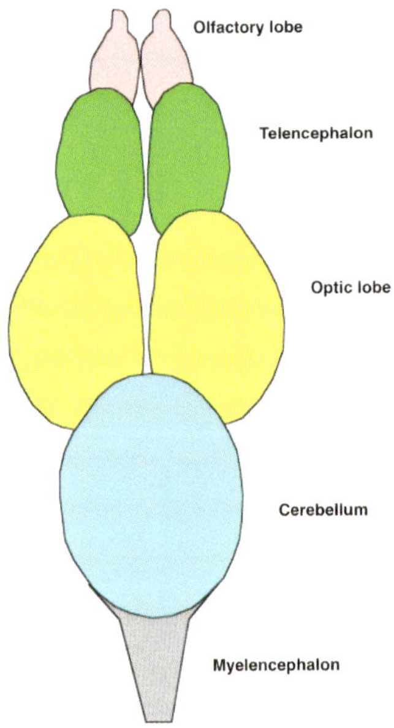

Dorsal view of the brain of the rainbow trout

Fish typically have quite small brains relative to body size compared with other vertebrates, typically one-fifteenth the brain mass of a similarly sized bird or mammal. However, some fish have relatively large brains, most notably mormyrids and sharks, which have brains about as massive relative to body weight as birds and marsupials.

Fish brains are divided into several regions. At the front are the olfactory lobes, a pair of structures that receive and process signals from the nostrils via the two olfactory nerves. The olfactory lobes are very large in fish that hunt primarily by smell, such as hagfish, sharks, and catfish. Behind the olfactory lobes is the two-lobed telencephalon, the structural equivalent to the cerebrum in higher vertebrates. In fish the telencephalon is concerned mostly with olfaction. Together these structures form the forebrain.

Connecting the forebrain to the midbrain is the diencephalon. The diencephalon performs functions associated with hormones and homeostasis. The pineal body lies just above the diencephalon. This structure detects light, maintains circadian rhythms, and controls color changes.

The midbrain (or mesencephalon) contains the two optic lobes. These are very large in species that hunt by sight, such as rainbow trout and cichlids.

The hindbrain (or metencephalon) is particularly involved in swimming and balance.

The cerebellum is a single-lobed structure that is typically the biggest part of the brain. Hagfish and lampreys have relatively small cerebellae, while the mormyrid cerebellum is massive and apparently involved in their electrical sense.

The brain stem (or myelencephalon) is the brain's posterior. As well as controlling some muscles and body organs, in bony fish at least, the brain stem governs respiration and osmoregulation.

Sense Organs

Most fish possess highly developed sense organs. Nearly all daylight fish have color vision that is at least as good as a human's. Many fish also have chemoreceptors that are responsible for extraordinary senses of taste and smell. Although they have ears, many fish may not hear very well. Most fish have sensitive receptors that form the lateral line system, which detects gentle currents and vibrations, and senses the motion of nearby fish and prey. Some fish, such as catfish and sharks, have the Ampullae of Lorenzini, organs that detect weak electric currents on the order of millivolt. Other fish, like the South American electric fishes Gymnotiformes, can produce weak electric currents, which they use in navigation and social communication.

Fish orient themselves using landmarks and may use mental maps based on multiple landmarks or symbols. Fish behavior in mazes reveals that they possess spatial memory and visual discrimination.

Vision

Vision is an important sensory system for most species of fish. Fish eyes are similar to those of terrestrial vertebrates like birds and mammals, but have a more spherical lens. Their retinas generally have both rods and cones (for scotopic and photopic vision), and most species have colour vision. Some fish can see ultraviolet and some can see polarized light. Amongst jawless fish, the lamprey has well-developed eyes, while the hagfish has only primitive eyespots. Fish vision shows adaptation to their visual environment, for example deep sea fishes have eyes suited to the dark environment.

Hearing

Hearing is an important sensory system for most species of fish. Fish sense sound using their lateral lines and their ears.

Capacity for Pain

Experiments done by William Tavolga provide evidence that fish have pain and fear responses. For instance, in Tavolga's experiments, toadfish grunted when electrically shocked and over time they came to grunt at the mere sight of an electrode.

In 2003, Scottish scientists at the University of Edinburgh and the Roslin Institute concluded that rainbow trout exhibit behaviors often associated with pain in other animals. Bee venom and acetic acid injected into the lips resulted in fish rocking their bodies and rubbing their lips along the sides and floors of their tanks, which the researchers concluded were attempts to relieve pain, similar to what mammals would do. Neurons fired in a pattern resembling human neuronal patterns.

Professor James D. Rose of the University of Wyoming claimed the study was flawed since it did not provide proof that fish possess "conscious awareness, particularly a kind of awareness that is meaningfully like ours". Rose argues that since fish brains are so different from human brains, fish are probably not conscious in the manner humans are, so that reactions similar to human reactions to pain instead have other causes. Rose had published a study a year earlier arguing that fish cannot feel pain because their brains lack a neocortex. However, animal behaviorist Temple Grandin argues that fish could still have consciousness without a neocortex because "different species can use different brain structures and systems to handle the same functions."

Animal welfare advocates raise concerns about the possible suffering of fish caused by angling. Some countries, such as Germany have banned specific types of fishing, and the British RSPCA now formally prosecutes individuals who are cruel to fish.

Muscular System

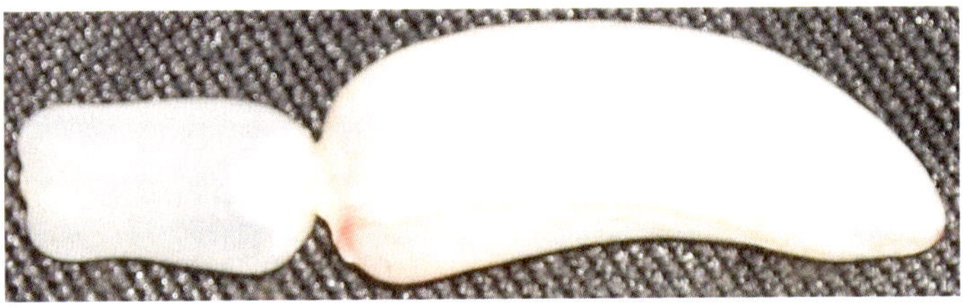

Swim bladder of a rudd (*Scardinius erythrophthalmus*)

Most fish move by alternately contracting paired sets of muscles on either side of the backbone. These contractions form S-shaped curves that move down the body. As each curve reaches the back fin, backward force is applied to the water, and in conjunction with the fins, moves the fish forward. The fish's fins function like an airplane's flaps. Fins also increase the tail's surface area, increasing speed. The streamlined body of the fish decreases the amount of friction from the water. Since body tissue is denser than water, fish must compensate for the difference or they will sink. Many bony fish have an internal organ called a swim bladder that adjusts their buoyancy through manipulation of gases.

Endothermy

Although most fish are exclusively ectothermic, there are exceptions. The only known bony fishes (infraclass Teleostei) that exhibit endothermy are in the suborder Scombroidei – which includes the billfishes, tunas, and the butterfly kingfish, a basal species of mackerel – and also the opah. The opah, a lampriform, was demonstrated in 2015 to utilize "whole-body endothermy", generating heat with its swimming muscles to warm its body while countercurrent exchange (as in respiration) minimizes heat loss. It is able to actively hunt prey such as squid and swim for long distances due to the ability to warm its entire body, including its heart, which is a trait typically found in only mammals and birds (in the form of homeothermy). In the cartilaginous fishes (class Chondrichthyes), sharks of the families Lamnidae (porbeagle, mackerel, salmon, and great white sharks) and Alopiidae (thresher sharks) exhibit endothermy. The degree of endothermy varies from the billfishes, which warm only their eyes and brain, to the bluefin tuna and the porbeagle shark, which maintain body temperatures in excess of 20 °C (68 °F) above ambient water temperatures.

Endothermy, though metabolically costly, is thought to provide advantages such as increased muscle strength, higher rates of central nervous system processing, and higher rates of digestion.

Reproductive System

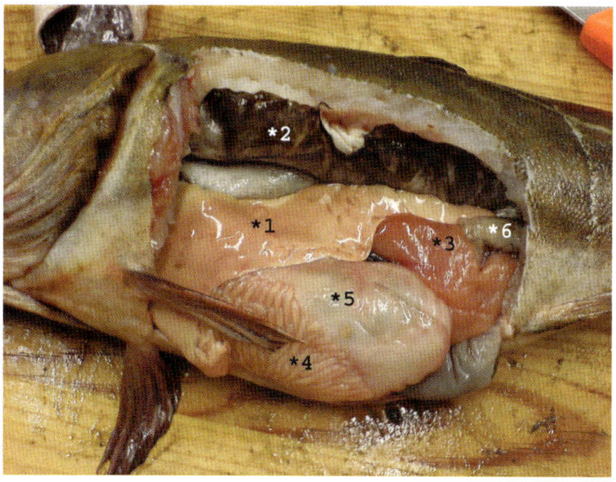

Organs: 1. Liver, 2. Gas bladder, 3. Roe, 4. Pyloric caeca, 5. Stomach, 6. Intestine

Fish reproductive organs include testicles and ovaries. In most species, gonads are paired organs of similar size, which can be partially or totally fused. There may also be a range of secondary organs that increase reproductive fitness.

In terms of spermatogonia distribution, the structure of teleosts testes has two types: in the most common, spermatogonia occur all along the seminiferous tubules, while in atherinomorph fish they are confined to the distal portion of these structures. Fish can present cystic or semi-cystic spermatogenesis in relation to the release phase of germ cells in cysts to the seminiferous tubules lumen.

Fish ovaries may be of three types: gymnovarian, secondary gymnovarian or cystovarian. In the first type, the oocytes are released directly into the coelomic cavity and then enter the ostium, then through the oviduct and are eliminated. Secondary gymnovarian ovaries shed ova into the coelom from which they go directly into the oviduct. In the third type, the oocytes are conveyed to the exterior through the oviduct. Gymnovaries are the primitive condition found in lungfish, sturgeon, and bowfin. Cystovaries characterize most teleosts, where the ovary lumen has continuity with the oviduct. Secondary gymnovaries are found in salmonids and a few other teleosts.

Oogonia development in teleosts fish varies according to the group, and the determination of oogenesis dynamics allows the understanding of maturation and fertilization processes. Changes in the nucleus, ooplasm, and the surrounding layers characterize the oocyte maturation process.

Postovulatory follicles are structures formed after oocyte release; they do not have endocrine function, present a wide irregular lumen, and are rapidly reabsorbed in a process involving the apoptosis of follicular cells. A degenerative process called follicular atresia reabsorbs vitellogenic oocytes not spawned. This process can also occur, but less frequently, in oocytes in other development stages.

Some fish, like the California sheephead, are hermaphrodites, having both testes and ovaries either at different phases in their life cycle or, as in hamlets, have them simultaneously.

Over 97% of all known fish are oviparous, that is, the eggs develop outside the mother's body. Examples of oviparous fish include salmon, goldfish, cichlids, tuna, and eels. In the majority of these species, fertilisation takes place outside the mother's body, with the male and female fish shedding their gametes into the surrounding water. However, a few oviparous fish practice internal fertilization, with the male using some sort of intromittent organ to deliver sperm into the genital opening of the female, most notably the oviparous sharks, such as the horn shark, and oviparous rays, such as skates. In these cases, the male is equipped with a pair of modified pelvic fins known as claspers.

Marine fish can produce high numbers of eggs which are often released into the open water column. The eggs have an average diameter of 1 millimetre (0.039 in).

Egg of lamprey

Egg of catshark (mermaids' purse)

Egg of bullhead shark

Egg of chimaera

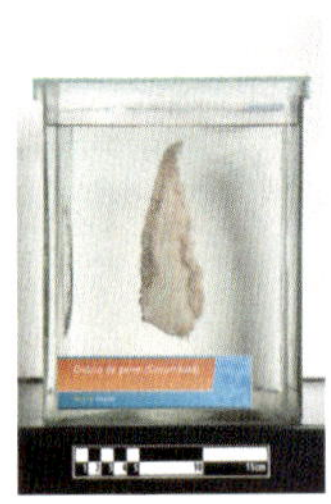

Ovary of fish (Corumbatá).

The newly hatched young of oviparous fish are called larvae. They are usually poorly formed, carry a large yolk sac (for nourishment), and are very different in appearance from juvenile and adult specimens. The larval period in oviparous fish is relatively short (usually only several weeks), and larvae rapidly grow and change appearance and structure (a process termed metamorphosis) to become juveniles. During this transition larvae must switch from their yolk sac to feeding on zooplankton prey, a process which depends on typically inadequate zooplankton density, starving many larvae.

In ovoviviparous fish the eggs develop inside the mother's body after internal fertilization but receive little or no nourishment directly from the mother, depending instead on the yolk. Each embryo develops in its own egg. Familiar examples of ovoviviparous fish include guppies, angel sharks, and coelacanths.

Some species of fish are viviparous. In such species the mother retains the eggs and nourishes the embryos. Typically, viviparous fish have a structure analogous to the placenta seen in mammals connecting the mother's blood supply with that of the embryo. Examples of viviparous fish include the surf-perches, splitfins, and lemon shark. Some viviparous fish exhibit oophagy, in which the developing embryos eat other eggs produced by the mother. This has been observed primarily among sharks, such as the shortfin mako and porbeagle, but is known for a few bony fish as well, such as the halfbeak *Nomorhamphus ebrardtii*. Intrauterine cannibalism is an even more unusual mode of vivipary, in which the largest embryos eat weaker and smaller siblings. This behavior is also most commonly found among sharks, such as the grey nurse shark, but has also been reported for *Nomorhamphus ebrardtii.*

Aquarists commonly refer to ovoviviparous and viviparous fish as livebearers.

Invertebrate

Invertebrates are animals without a backbone or bony skeleton.

They range in size from microscopic mites and almost invisible flies to giant squid with soccer-ball-size eyes.

This is by far the largest group in the animal kingdom: 97 percent of all animals are invertebrates. So far, 1.25 million species have been described, most of which are insects, and there are millions more to be discovered. The total number of invertebrate species could be 5, 10, or even 30 million, com- pared to just 60,000 vertebrates.

One reason for the success of invertebrates is how quickly they reproduce. Sponges and corals, for example, produce both eggs and sperm. Social insects such as ants and bees lay eggs that can develop without fertilization—they become the workers.

Insects in particular are successful because they are so adaptable. They are opportunistic eaters, feeding on plants, animals, and decaying organic material. They are able to survive in extreme environments, including very hot, dry habitats. And many can fly—either to escape predators or to find new sources of food, water, and shelter.

Like vertebrates, invertebrates are classified based upon their body structure, life cycle, and evolutionary history.

Characteristics

The trait that is common to all invertebrates is the absence of a vertebral column (backbone): this creates a distinction between invertebrates and vertebrates. The distinction is one of convenience only; it is not based on any clear biologically homologous trait, any more than the common trait of having wings functionally unites insects, bats, and birds, or than not having wings unites tortoises, snails and sponges. Being animals, invertebrates are heterotrophs, and require sustenance in the form of the consumption of other organisms. With a few exceptions, such as the Porifera, invertebrates generally have bodies composed of differentiated tissues. There is also typically a digestive chamber with one or two openings to the exterior.

Morphology and Symmetry

The body plans of most multicellular organisms exhibit some form of symmetry, whether radial, bilateral, or spherical. A minority, however, exhibit no symmetry. One example of asymmetric invertebrates include all gastropod species. This is easily seen in snails and sea snails, which have helical shells. Slugs appear externally symmetrical, but their pneumostome (breathing hole) is located on the right side. Other gastropods develop external asymmetry, such as Glaucus atlanticus that develops asymmetrical cerata as they mature. The origin of gastropod asymmetry is a subject of scientific debate.

Other examples of asymmetry are found in fiddler crabs and hermit crabs. They often have one claw much larger than the other. If a male fiddler loses its large claw, it will grow another on the opposite side after moulting. Sessile animals such as sponges are asymmetrical alongside coral colonies (with the exception of the individual polyps that exhibit radial symmetry); alpheidae claws that lack pincers; and some copepods, polyopisthocotyleans, and monogeneans which parasitize by attachment or residency within the gill chamber of their fish hosts).

Nervous System

Neurons differ in invertebrates from mammalian cells. Invertebrates cells fire in response to similar stimuli as mammals, such as tissue trauma, high temperature, or changes in pH. The first invertebrate in which a neuron cell was identified was the medicinal leech, *Hirudo medicinalis*.

Learning and memory using nociceptors in the sea hare, *Aplysia* has been described. Mollusk neurons are able to detect increasing pressures and tissue trauma.

Neurons have been identified in a wide range of invertebrate species, including annelids, molluscs, nematodes and arthropods.

Respiratory System

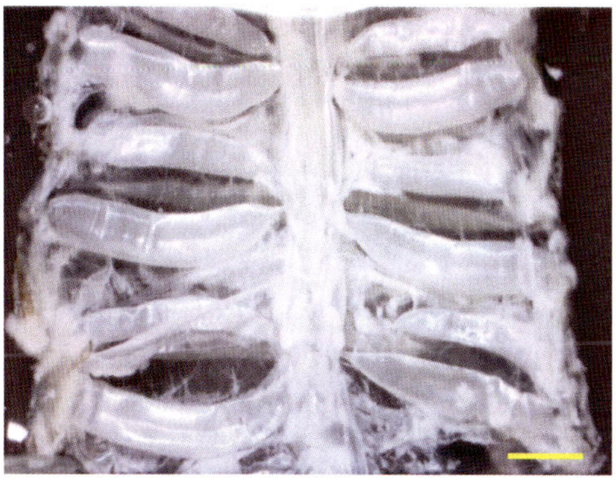

Tracheal system of dissected cockroach. The largest tracheae run across the width of the body of the cockroach and are horizontal in this image.

The tracheal system branches into progressively smaller tubes, here supplying the crop of the cockroach.

One type of invertebrate respiriatory system is the open respiratory system composed of spiracles, tracheae, and tracheoles that terrestrial arthropods have to transport metabolic gases to and from tissues. The distribution of spiracles can vary greatly among the many orders of insects, but in general each segment of the body can have only one pair of spiracles, each of which connects to an atrium and has a relatively large tracheal tube behind it. The tracheae are invaginations of the cuticular exoskeleton that branch (anastomose) throughout the body with diameters from only a few micrometres up to 0.8 mm. The smallest tubes, tracheoles, penetrate cells and serve as sites of diffusion for water, oxygen, and carbon dioxide. Gas may be conducted through the respiratory system by means of active ventilation or passive diffusion. Unlike vertebrates, insects do not generally carry oxygen in their haemolymph.

A tracheal tube may contain ridge-like circumferential rings of taenidia in various geometries such as loops or helices. In the head, thorax, or abdomen, tracheae may also be connected to air sacs. Many insects, such as grasshoppers and bees, which actively pump the air sacs in their abdomen, are able to control the flow of air through their body. In some aquatic insects, the tracheae exchange gas through the body wall directly, in the form of a gill, or function essentially as normal, via a plastron. Note that despite being internal, the tracheae of arthropods are shed during moulting (ecdysis).

Reproduction

Like vertebrates, most invertebrates reproduce at least partly through sexual reproduction. They produce specialized reproductive cells that undergo meiosis to produce smaller, motile spermatozoa or larger, non-motile ova. These fuse to form zygotes, which develop into new individuals. Others are capable of asexual reproduction, or sometimes, both methods of reproduction.

Social Interaction

Social behavior is widespread in invertebrates, including cockroaches, termites, aphids, thrips, ants, bees, Passalidae, Acari, spiders, and more. Social interaction is particularly salient in eusocial species but applies to other invertebrates as well.

Insects recognize information transmitted by other insects.

Phyla

The fossil coral *Cladocora* from the Pliocene of Cyprus

The term invertebrates covers several phyla. One of these are the sponges (Porifera). They were long thought to have diverged from other animals early. They lack the complex organization found in most other phyla. Their cells are differentiated, but in most cases not organized into distinct tissues. Sponges typically feed by drawing in water through pores. Some speculate that sponges are not so primitive, but may instead be secondarily simplified. The Ctenophora and the Cnidaria, which includes sea anemones, corals, and jellyfish, are radially symmetric and have digestive chambers with a single opening, which serves as both the mouth and the anus. Both have distinct tissues, but they are not organized into organs. There are only two main germ layers, the ectoderm and endoderm, with only scattered cells between them. As such, they are sometimes called diploblastic.

The Echinodermata are radially symmetric and exclusively marine, including starfish (Asteroidea), sea urchins, (Echinoidea), brittle stars (Ophiuroidea), sea cucumbers (Holothuroidea) and feather stars (Crinoidea).

The largest animal phylum is also included within invertebrates: the Arthropoda, including insects, spiders, crabs, and their kin. All these organisms have a body divided into repeating segments, typically with paired appendages. In addition, they possess a hardened exoskeleton that is periodically shed during growth. Two smaller phyla, the Onychophora and Tardigrada, are close relatives of the arthropods and share these traits. The Nematoda or roundworms, are perhaps the second largest animal phylum, and are also invertebrates. Roundworms are typically microscopic, and occur in nearly every environment where there is water. A number are important parasites. Smaller phyla related to them are the Kinorhyncha, Priapulida, and Loricifera. These groups have a reduced coelom, called a pseudocoelom. Other invertebrates include the Nemertea or ribbon worms, and the Sipuncula.

Another phylum is Platyhelminthes, the flatworms. These were originally considered primitive, but it now appears they developed from more complex ancestors. Flatworms are acoelomates, lacking a body cavity, as are their closest relatives, the microscopic Gastrotricha. The Rotifera or rotifers, are common in aqueous environments. Invertebrates also include the Acanthocephala or spiny-headed worms, the Gnathostomulida, Micrognathozoa, and the Cycliophora.

Also included are two of the most successful animal phyla, the Mollusca and Annelida. The former, which is the second-largest animal phylum by number of described species, includes animals such as snails, clams, and squids, and the latter comprises the segmented worms, such as earthworms and leeches. These two groups have long been considered close relatives because of the common presence of trochophore larvae, but the annelids were considered closer to the arthropods because they are both segmented. Now, this is generally considered convergent evolution, owing to many morphological and genetic differences between the two phyla.

Among lesser phyla of invertebrates are the Hemichordata, or acorn worms, and the Chaetognatha, or arrow worms. Other phyla include Acoelomorpha, Brachiopoda, Bryozoa, Entoprocta, Phoronida, and Xenoturbellida.

Classifying an Invertebrate

In our deserts, the invertebrate subkingdom includes phyla such as arthropods (insects, centipedes, spiders, scorpions, desert shrimp and many others), mollusks (snails) and annelids (earthworms). In the desert, as well as across the world, arthropods, measured in terms of abundance and diversity, rank at the top of all the animal phyla. An elegant insect, the monarch butterfly, serves as example of how the classification system works for the invertebrates.

At the phylum level, the monarch belongs to the arthropods, which share several physical characteristics. According to Barbara Terkanian, "A Vertebrate Looks At Arthropods," *A Natural History of the Sonoran Desert,* the arthropods have jointed legs, and they have external skeletons, or exoskeletal material, that includes "eyes, mouthparts, antennae, body, legs, the fore and hind sections of the digestive tract, and some respiratory surfaces. Regions of flexible, unhardened exoskeleton serve as joints between neighboring segments." The body cavity contains the digestive, circulatory, nervous and reproductive systems.

At the class level, the monarch has membership in the insect group, which comprises the overwhelming majority of the arthropods. The insects have several distinguishing physical characteristics, including three-part bodies, six legs (three pairs), compound eyes and two antennae. The class, called *Insecta,* includes three subclasses, according to Kendall Bioresearch Services Internet site. The first consists of insects that have never had wings throughout their evolutionary history. The young resemble the adults. The second subclass consists of insects that have wings at present or had them at some point during their evolutionary history. The nymphs resemble the adults. The third subclass consists of insects that have wings at present or had them at some point during their evolutionary history. The young take the form of larvae that change into adults during a non-feeding metamorphosis. The first subclass consists of four orders, including, for example, bristletails and springs tails. The second subclass has 16 orders, including, for instance, dragonflies; crickets, grasshoppers and locusts; termites and sucking lice. The third subclass has nine orders, comprising insects such as beetles; fleas; bees, wasps and ants; and the butterflies and moths.

At the order level, the monarch belongs to butterflies and moths, called *Lepidoptera,* which rank high among the most intriguing and conspicuous insect orders in the Southwest. They have two pairs of membranous, scaled and often brightly colored wings. Typically they have large eyes, long antennae and a long sucking tube (which the insect coils beneath its head when not feeding). The larvae, or caterpillars, all have silk glands that they use for spinning their cocoons. Their order contains well over 100 families.

At the family level, the monarch is the star of the milkweed butterflies, called *Danaidae,* which are among the best known in our deserts (as well as across the country). The milkweed butterflies usually have goldish wings trimmed in black, according to Donald J. Borror and Richard E. White, *A Field Guide to the Insects of America North of Mexico.* Their caterpillars feed on milkweed leaves, which invest both larvae and adults with a bitter and toxic taste that discourages predators.

At the genera level, the monarch is one of a mere handful of closely related species collectively called *Dannaus.* These species show apparently common evolutionary origins in their caterpillars, which share similar spots and smooth skin texture on their abdomens.

At the species level, the monarch is called *plexippus.* It is a medium sized butterfly, measuring about 3 inches from wingtip to wingtip. Its body is about one inch long. Its four wings are generally a field of yellow, orange or gold, with veins of black running through them. A band of black, thickest at the front, rings the wings, and the body is black as well. This black band is usually speckled with white spots, larger at the front and smaller at the back.

The monarch, the aristocrat of the butterfly and moth world, bears the scientific name of *Dannaus plexippus.* In summary, it fits into the Linnaeus classification scheme as follows:

Classification Level	**Scientific Name**
Kingdom	Animalia
Subkingdom	Invertebrate
Phylum	*Arthropoda*
Class	*Insecta*
Order	*Lepidoptera*
Family	*Danaidae*
Genus	*Dannaus*
Species	Plexippus
Common Name	Monarch Butterfly
Scientific Name	*Dannaus plexippus*

In one of the most compelling of nature's mysteries, the monarchs, which have a nationwide distribution, begin annual epic migrations by the millions during the autumn. Those from our Southwestern states to the Northwest migrate to southern California. Those from the Midwest to the eastern seaboard migrate across the southern Chihuahuan Desert to the mountains of central Mexico, covering thousands of miles. Their journey is closely watched along the entire route. Their arrival in Mexico calls for celebration: "The butterflies have arrived in Saltillo (in the southeastern corner of the Chihuahuan Desert)!" one Mexican woman reported to the Monarch Butterfly Internet site in 1999. "Since early this morning (October 12), we have welcomed the arrival of the monarchs throughout the city, there have been hundreds of monarchs gliding through from 3 to 50 meters in height"

The butterfly, like the moth, is distinguished by its membranous, scaled and often brightly colored wings.

Main Invertebrate Group

Porifera

The Porifera (sponges) differ from other animals in that they do not exhibit true tissue development but rather have a cellular level of organisation. They are simple in body form, having a sac-like body which is perforated by many pores, yet complex at the cellular level having several types of cells. The outer layer of the cells wall contains flattened epidermal cells, some of which have contractile fibers; the middle layer is a semifluid matrix with wandering amoebocytes; and the inner layer is composed of flagellated cells called collar cells (or choanocytes) that look like protozoans. One could thus argue that sponges are composed of colonies of protozoans living in a symbiotic relationship with each other.

All sponges are sessile, suspension feeding animals with radial or no symmetry. Only their sperm and larvae are motile and once a larva settles it attaches itself to a substrate and develops into an adult. Sponges are filter feeders and are the only animals in which digestion occurs within cells. They may be as tiny as a finger nail to as large as a person. Even though many evolutionary biologists believe them to be an evolutionary dead end, sponges are one of nature's success stories.

There are over 5000 species of sponges known globally and are generally classified according to their skeletal structures. Their body shape is maintained by possessing spicules, made of calcium carbonate or silicate, or collagenous fibers known as spongin. The class Calcarea possesses calcium carbonate spicules and members thereof are exclusively marine. The class Hexactinellida or glass sponges, primarily found in deep marine habitats, possesses spicules composed of silica. The class Demospongiae, found in marine, brackish and fresh water environments, possesses both spicules and spongin. At present 314 Poriferan species are known from South Africa. This species diversity compares well with most regions that have well documented sponge faunas but is unexceptional when compared to Australia (2426 species), although this later comparison cannot strictly be made due to the vast differences in sampling effort, area and habitat diversity.

Basic Structure

Cell Types

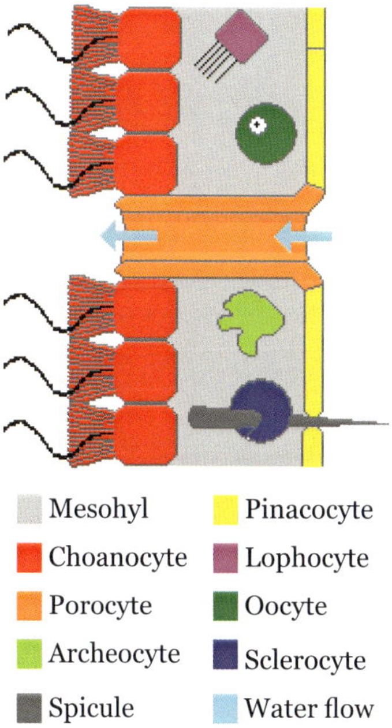

Main cell types of Porifera

A sponge's body is hollow and is held in shape by the mesohyl, a jelly-like substance made mainly of collagen and reinforced by a dense network of fibers also made of collagen. The inner surface is covered with choanocytes, cells with cylindrical or conical collars surrounding one flagellum per choanocyte. The wave-like motion of the whip-like flagella drives water through the sponge's body. All sponges have ostia, channels leading to the interior through the mesohyl, and in most sponges these are controlled by tube-like porocytes that form closable inlet valves. Pinacocytes, plate-like cells, form a single-layered external skin over all other parts of the mesohyl that are not covered by choanocytes, and the pinacocytes also digest food particles that are too large to enter the ostia, while those at the base of the animal are responsible for anchoring it.

Other types of cell live and move within the mesohyl:

- Lophocytes are amoeba-like cells that move slowly through the mesohyl and secrete collagen fibres.
- Collencytes are another type of collagen-producing cell.
- Rhabdiferous cells secrete polysaccharides that also form part of the mesohyl.
- Oocytes and spermatocytes are reproductive cells.
- Sclerocytes secrete the mineralized spicules ("little spines") that form the skeletons of many sponges and in some species provide some defense against predators.
- In addition to or instead of sclerocytes, demosponges have spongocytes that secrete a form of collagen that polymerizes into spongin, a thick fibrous material that stiffens the mesohyl.
- Myocytes ("muscle cells") conduct signals and cause parts of the animal to contract.
- "Grey cells" act as sponges' equivalent of an immune system.
- Archaeocytes (or amoebocytes) are amoeba-like cells that are totipotent, in other words each is capable of transformation into any other type of cell. They also have important roles in feeding and in clearing debris that block the ostia.

Glass Sponges' Syncytia

Glass sponges present a distinctive variation on this basic plan. Their spicules, which are made of silica, form a scaffolding-like framework between whose rods the living tissue is suspended like a cobweb that contains most of the cell types. This tissue is a syncytium that in some ways behaves like many cells that share a single external membrane, and in others like a single cell with multiple nuclei. The mesohyl is absent or minimal. The syncytium's cytoplasm, the soupy fluid that fills the interiors of cells, is organized into "rivers" that transport nuclei, organelles ("organs" within cells) and other substances. Instead of choanocytes, they have further syncytia, known as choanosyncytia, which form bell-shaped chambers where water enters via perforations. The insides of these chambers are lined with "collar bodies", each consisting of a collar and flagellum but without a nucleus of its own. The motion of the flagella sucks water through passages in the "cobweb" and expels it via the open ends of the bell-shaped chambers.

Water flow
Main syncitium
Spicules
Choanosyncitium and collar bodies showing interior

The glass sponge *Euplectella*

Some types of cells have a single nucleus and membrane each, but are connected to other single-nucleus cells and to the main syncytium by "bridges" made of cytoplasm. The sclerocytes that build spicules have multiple nuclei, and in glass sponge larvae they are connected to other tissues by cytoplasm bridges; such connections between sclerocytes have not so far been found in adults, but this may simply reflect the difficulty of investigating such small-scale features. The bridges are controlled by "plugged junctions" that apparently permit some substances to pass while blocking others.

Water Flow and Body Structures

Most sponges work rather like chimneys: they take in water at the bottom and eject it from the osculum ("little mouth") at the top. Since ambient currents are faster at the top, the suction effect that they produce by Bernoulli's principle does some of the work for free. Sponges can control the water flow by various combinations of wholly or partially closing the osculum and ostia (the intake pores) and varying the beat of the flagella, and may shut it down if there is a lot of sand or silt in the water.

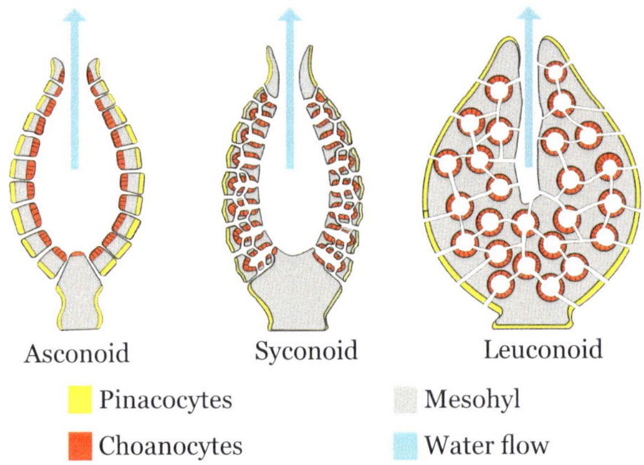

Porifera body structures

Although the layers of pinacocytes and choanocytes resemble the epithelia of more complex animals, they are not bound tightly by cell-to-cell connections or a basal lamina (thin fibrous sheet underneath). The flexibility of these layers and re-modeling of the mesohyl by lophocytes allow the animals to adjust their shapes throughout their lives to take maximum advantage of local water currents.

The simplest body structure in sponges is a tube or vase shape known as "asconoid", but this severely limits the size of the animal. The body structure is characterized by a stalk-like spongocoel surrounded by a single layer of choanocytes. If it is simply scaled up, the ratio of its volume to surface area increases, because surface increases as the square of length or width while volume increases proportionally to the cube. The amount of tissue that needs food and oxygen is determined by the volume, but the pumping capacity that supplies food and oxygen depends on the area covered by choanocytes. Asconoid sponges seldom exceed 1 mm (0.039 in) in diameter.

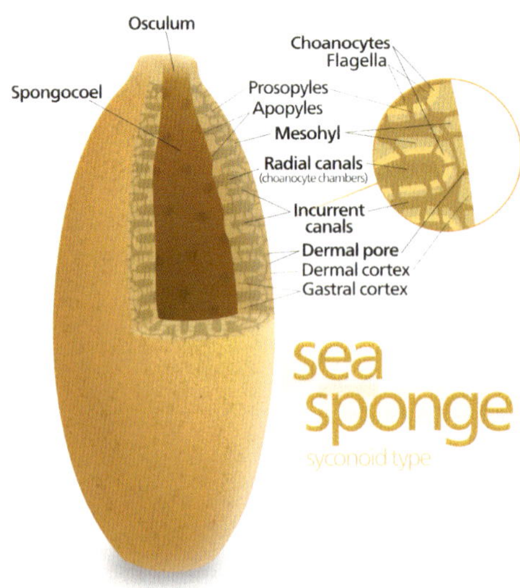

Diagram of a syconoid sponge

Some sponges overcome this limitation by adopting the "syconoid" structure, in which the body wall is pleated. The inner pockets of the pleats are lined with choanocytes, which connect to the outer pockets of the pleats by ostia. This increase in the number of choanocytes and hence in pumping capacity enables syconoid sponges to grow up to a few centimeters in diameter.

The "leuconoid" pattern boosts pumping capacity further by filling the interior almost completely with mesohyl that contains a network of chambers lined with choanocytes and connected to each other and to the water intakes and outlet by tubes. Leuconid sponges grow to over 1 m (3.3 ft) in diameter, and the fact that growth in any direction increases the number of choanocyte chambers enables them to take a wider range of forms, for example "encrusting" sponges whose shapes follow those of the surfaces to which they attach. All freshwater and most shallow-water marine sponges have leuconid bodies. The networks of water passages in glass sponges are similar to the leuconid structure. In all three types of structure the cross-section area of the choanocyte-lined regions is much greater than that of the intake and outlet channels. This makes the flow slower near the choanocytes and thus makes it easier for them to trap food particles. For example, in *Leuconia*, a small leuconoid sponge about 10 centimetres (3.9 in) tall and 1 centimetre (0.39 in) in diameter, water enters each of more than 80,000 intake canals at 6 cm per *minute*. However, because *Leuconia* has more than 2 million flagellated chambers whose combined diameter is much greater than that of the canals, water flow through chambers slows to 3.6 cm per *hour*, making it easy for choanocytes to capture food. All the water is expelled through a single osculum at about 8.5 cm per *second*, fast enough to carry waste products some distance away.

Skeleton

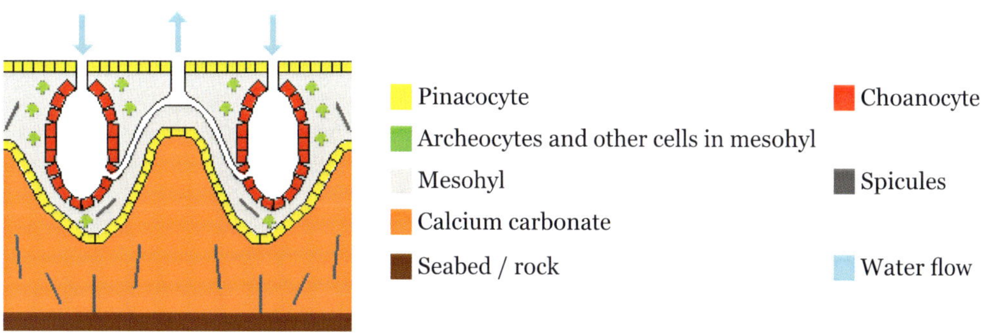

Sponge with calcium carbonate skeleton

In zoology a skeleton is any fairly rigid structure of an animal, irrespective of whether it has joints and irrespective of whether it is biomineralized. The mesohyl functions as an endoskeleton in most sponges, and is the only skeleton in soft sponges that encrust hard surfaces such as rocks. More commonly the mesohyl is stiffened by mineral spicules, by spongin fibers or both. Spicules may be made of silica or calcium carbonate, and vary in shape from simple rods to three-dimensional "stars" with up to six rays. Spicules are produced by sclerocyte cells, and may be separate, connected by joints, or fused.

Some sponges also secrete exoskeletons that lie completely outside their organic components. For example, sclerosponges ("hard sponges") have massive calcium carbonate exoskeletons over which the organic matter forms a thin layer with choanocyte chambers in pits in the mineral. These exoskeletons are secreted by the pinacocytes that form the animals' skins.

Vital Functions

Movement

Spongia officinalis, "the kitchen sponge", is dark grey when alive.

Although adult sponges are fundamentally sessile animals, some marine and freshwater species can move across the sea bed at speeds of 1–4 mm (0.039–0.157 in) per day, as a result of amoeba-like movements of pinacocytes and other cells. A few species can contract their whole bodies, and many can close their oscula and ostia. Juveniles drift or swim freely, while adults are stationary.

Respiration, Feeding and Excretion

Sponges do not have distinct circulatory, respiratory, digestive, and excretory systems – instead the water flow system supports all these functions. They filter food particles out of the water flowing through them. Particles larger than 50 micrometers cannot enter the ostia and pinacocytes consume them by phagocytosis (engulfing and internal digestion). Particles from 0.5 µm to 50 µm are trapped in the ostia, which taper from the outer to inner ends. These particles are consumed by pinacocytes or by archaeocytes which partially extrude themselves through the walls of the ostia. Bacteria-sized particles, below 0.5 micrometers, pass through the ostia and are caught and consumed by choanocytes. Since the smallest particles are by far the most common, choanocytes typically capture 80% of a sponge's food supply. Archaeocytes transport food packaged in vesicles from cells that directly digest food to those that do not. At least one species of sponge has internal fibers that function as tracks for use by nutrient-carrying archaeocytes, and these tracks also move inert objects.

Euplectella aspergillum, a glass sponge known as "Venus' flower basket"

It used to be claimed that glass sponges could live on nutrients dissolved in sea water and were very averse to silt. However, a study in 2007 found no evidence of this and concluded that they extract bacteria and other micro-organisms from water very efficiently (about 79%) and process suspended sediment grains to extract such prey. Collar bodies digest food and distribute it wrapped in vesicles that are transported by dynein "motor" molecules along bundles of microtubules that run throughout the syncytium.

Sponges' cells absorb oxygen by diffusion from water into cells as water flows through body, into which carbon dioxide and other soluble waste products such as ammonia also diffuse. Archeocytes remove mineral particles that threaten to block the ostia, transport them through the mesohyl and generally dump them into the outgoing water current, although some species incorporate them into their skeletons.

Carnivorous Sponges

A few species that live in waters where the supply of food particles is very poor prey on crustaceans and other small animals. So far only 137 species have been discovered. Most belong to the family Cladorhizidae, but a few members of the Guitarridae and Esperiopsidae are also carnivores. In most cases little is known about how they actually capture prey, although some species are thought to use either sticky threads or hooked spicules. Most carnivorous sponges live in deep waters, up to 8,840 m (5.49 mi), and the development of deep-ocean exploration techniques is expected to lead to the discovery of several more. However, one species has been found in Mediterranean caves at depths of 17–23 m (56–75 ft), alongside the more usual filter feeding sponges. The cave-dwelling predators capture crustaceans under 1 mm (0.039 in) long by entangling them with fine threads, digest them by enveloping them with further threads over the course of a few days, and then return to their normal shape; there is no evidence that they use venom.

Most known carnivorous sponges have completely lost the water flow system and choanocytes. However, the genus *Chondrocladia* uses a highly modified water flow system to inflate balloon-like structures that are used for capturing prey.

Endosymbionts

Freshwater sponges often host green algae as endosymbionts within archaeocytes and other cells, and benefit from nutrients produced by the algae. Many marine species host other photosynthesizing organisms, most commonly cyanobacteria but in some cases dinoflagellates. Symbiotic cyanobacteria may form a third of the total mass of living tissue in some sponges, and some sponges gain 48% to 80% of their energy supply from these micro-organisms. In 2008 a University of Stuttgart team reported that spicules made of silica conduct light into the mesohyl, where the photosynthesizing endosymbionts live. Sponges that host photosynthesizing organisms are most common in waters with relatively poor supplies of food particles, and often have leafy shapes that maximize the amount of sunlight they collect.

A recently discovered carnivorous sponge that lives near hydrothermal vents hosts methane-eating bacteria, and digests some of them.

Immune System

Sponges do not have the complex immune systems of most other animals. However, they reject grafts from other species but accept them from other members of their own species. In a few marine species, gray cells play the leading role in rejection of foreign material. When invaded, they produce a chemical that stops movement of other cells in the affected area, thus preventing the intruder from using the sponge's internal transport systems. If the intrusion persists, the grey cells concentrate in the area and release toxins that kill all cells in the area. The "immune" system can stay in this activated state for up to three weeks.

Reproduction

Asexual

Sponges have three asexual methods of reproduction: after fragmentation; by budding; and by producing gemmules. Fragments of sponges may be detached by currents or waves. They use the mobility of their pinacocytes and choanocytes and reshaping of the mesohyl to re-attach themselves to a suitable surface and then rebuild themselves as small but functional sponges over the course of several days. The same capabilities enable sponges that have been squeezed through a fine cloth to regenerate. A sponge fragment can only regenerate if it contains both collencytes to produce mesohyl and archeocytes to produce all the other cell types. A very few species reproduce by budding.

The freshwater sponge *Spongilla lacustris*

Gemmules are "survival pods" which a few marine sponges and many freshwater species produce by the thousands when dying and which some, mainly freshwater species, regularly produce in autumn. Spongocytes make gemmules by wrapping shells of spongin, often reinforced with spicules, round clusters of archeocytes that are full of nutrients. Freshwater gemmules may also include phytosynthesizing symbionts. The gemmules then become dormant, and in this state can survive cold, drying out, lack of oxygen and extreme variations in salinity. Freshwater gemmules often do not revive until the temperature drops, stays cold for a few months and then reaches a near-"normal" level. When a gemmule germinates, the archeocytes round the outside of the cluster transform into pinacocytes, a membrane over a pore in the shell bursts, the cluster of cells slowly emerges, and most of the remaining archeocytes transform into other cell types needed to make a functioning sponge. Gemmules from the same species but different individuals can join forces to form one sponge. Some gemmules are retained within the parent sponge, and in spring it can be difficult to tell whether an old sponge has revived or been "recolonized" by its own gemmules.

Sexual

Most sponges are hermaphrodites (function as both sexes simultaneously), although sponges have no gonads (reproductive organs). Sperm are produced by choanocytes or entire choanocyte chambers that sink into the mesohyl and form spermatic cysts while eggs are formed by transformation of archeocytes, or of choanocytes in some species. Each egg generally acquires a yolk by consuming "nurse cells". During spawning, sperm burst out of their cysts and are expelled via the osculum. If they contact another sponge of the same species, the water flow carries them to choanocytes that engulf them but, instead of digesting them, metamorphose to an ameboid form and carry the sperm through the mesohyl to eggs, which in most cases engulf the carrier and its cargo.

A few species release fertilized eggs into the water, but most retain the eggs until they hatch. There are four types of larvae, but all are balls of cells with an outer layer of cells whose flagellae or cilia enable the larvae to move. After swimming for a few days the larvae sink and crawl until they find a place to settle. Most of the cells transform into archeocytes and then into the types appropriate for their locations in a miniature adult sponge.

Glass sponge embryos start by dividing into separate cells, but once 32 cells have formed they rapidly transform into larvae that externally are ovoid with a band of cilia round the middle that they use for movement, but internally have the typical glass sponge structure of spicules with a cobweb-like main syncitium draped around and between them and choanosyncytia with multiple collar bodies in the center. The larvae then leave their parents' bodies.

Life Cycle

Sponges in temperate regions live for at most a few years, but some tropical species and perhaps some deep-ocean ones may live for 200 years or more. Some calcified demosponges grow by only 0.2 mm (0.0079 in) per year and, if that rate is constant, specimens 1 m (3.3 ft) wide must be about 5,000 years old. Some sponges start sexual reproduction when only a few weeks old, while others wait until they are several years old.

Coordination of Activities

Adult sponges lack neurons or any other kind of nervous tissue. However, most species have the ability to perform movements that are coordinated all over their bodies, mainly contractions of the pinacocytes, squeezing the water channels and thus expelling excess sediment and other substances that may cause blockages. Some species can contract the osculum independently of the rest of the body. Sponges may also contract in order to reduce the area that is vulnerable to attack by predators. In cases where two sponges are fused, for example if there is a large but still unseparated bud, these contraction waves slowly become coordinated in both of the "Siamese twins". The coordinating mechanism is unknown, but may involve chemicals similar to neurotransmitters. However, glass sponges rapidly transmit electrical impulses through all parts of the syncytium, and use this to halt the motion of their flagella if the incoming water contains toxins or excessive sediment. Myocytes are thought to be responsible for closing the osculum and for transmitting signals between different parts of the body.

Sponges contain genes very similar to those that contain the "recipe" for the post-synaptic density, an important signal-receiving structure in the neurons of all other animals. However, in sponges these genes are only activated in "flask cells" that appear only in larvae and may provide some sensory capability while the larvae are swimming. This raises questions about whether flask cells represent the predecessors of true neurons or are evidence that sponges' ancestors had true neurons but lost them as they adapted to a sessile lifestyle.

Cnidaria

The Cnidaria were previously known as Coelenterata, which sometimes included the Ctenophora (now considered a separate phylum). The Cnidaria encompasses corals, sea anemones, jelly-fish and hydroids. The systematics of the Cnidaria has undergone some major changes over the past 20 years as molecular data have become available and phylogenctic analyses have improved. The changes in taxonomy are especially prominent within the class Hydrozoa, but name changes at genus and family levels are numerous, and lists of synonyms are often long. Old names continue to appear in the literature, and this lack of stability is particularly problematic in management contexts, whether for conservation or biopollution assessments.

There are three major classes: Hydrozoa, Scyphozoa and Anthozoa. The Hydrozoa are the often highly branched, colonial hydroids as well as the fire corals, *Millepora* spp. and the pelagic, colonial Siphonophora. Many hydrozoans have a free swimming medusa stage, which is in fact the sexually reproducing stage rather than a larval stage. It is within this group that the majority of introduced species are found. This is most likely because the hydroid polyp stage can be part of the fouling community on ships' hulls and other man-made structures. The Scyphozoa are the majority of the large jelly-fish, such as the common moon jelly. Many species in this group have been known to form blooms, but so far no introduced species occur in Nordic waters. Anthozoa comprises the corals and sea anemones. Only deep water corals occur in Nordic waters, and no introduced species have been identified. For sea anemones some introduced species occur very close to our borders and/or have occasionally been found in Nordic waters. Anthozoans do not have a free swimming medusa stage; only the ciliated planula larvae are planktonic. In tropical waters the highly toxic box-jellies or sea-wasps, Cubozoa often occur in dangerous densities.

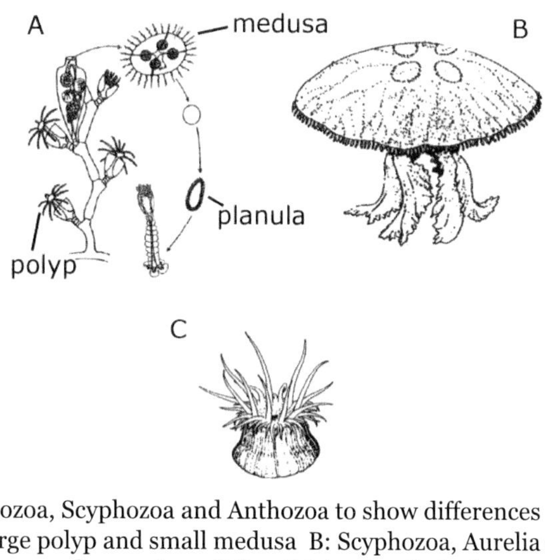

Drawings of Hydrozoa, Scyphozoa and Anthozoa to show differences A: Hydrozoa, Obelia sp. with large polyp and small medusa B: Scyphozoa, Aurelia sp. with large medusa and small (benthic) polyp C: Anthozoa, Djadumene lineala with no medusa and bilateral symmetry

Description

Basic Body Forms

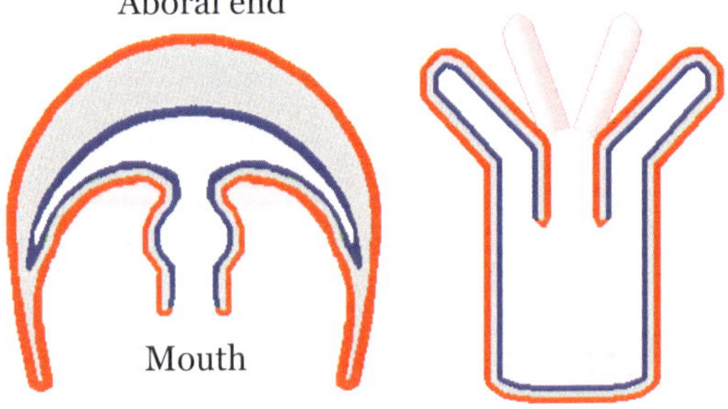

Medusa (left) and polyp (right)

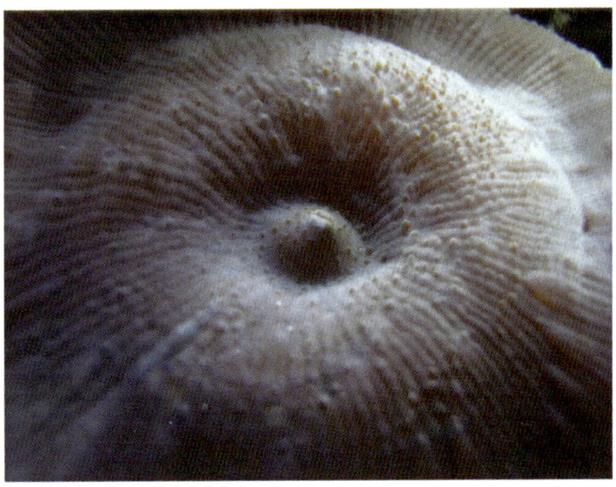

Oral end of actinodiscus polyp, with close-up of the mouth

Most adult cnidarians appear as either swimming medusae or sessile polyps, and many hydrozoan species are known to alternate between the two forms. Both are radially symmetrical, like a wheel and a tube respectively. Since these animals have no heads, their ends are described as "oral" (nearest the mouth) and "aboral" (furthest from the mouth). Most have fringes of tentacles equipped with cnidocytes around their edges, and medusae generally have an inner ring of tentacles around the mouth. Some hydroids may consist of colonies of zooids that serve different purposes, such as defense, reproduction and catching prey. The mesoglea of polyps is usually thin and often soft, but that of medusae is usually thick and springy, so that it returns to its original shape after muscles around the edge have contracted to squeeze water out, enabling medusae to swim by a sort of jet propulsion.

Skeletons

In medusae the only supporting structure is the mesoglea. *Hydra* and most sea anemones close their mouths when they are not feeding, and the water in the digestive cavity then acts as a hydrostatic skeleton, rather like a water-filled balloon. Other polyps such as *Tubularia* use columns of water-filled cells for support. Sea pens stiffen the mesoglea with calcium carbonate spicules and tough fibrous proteins, rather like sponges.

In some colonial polyps, a chitinous periderm gives support and some protection to the connecting sections and to the lower parts of individual polyps. Stony corals secrete massive calcium carbonate exoskeletons. A few polyps collect materials such as sand grains and shell fragments, which they attach to their outsides. Some colonial sea anemones stiffen the mesoglea with sediment particles.

Main Cell Layers

Cnidaria are diploblastic animals; in other words, they have two main cell layers, while more complex animals are triploblasts having three main layers. The two main cell layers of cnidarians form epithelia that are mostly one cell thick, and are attached to a fibrous basement membrane, which they secrete. They also secrete the jelly-like mesoglea that separates the layers. The layer that faces outwards, known as the ectoderm ("outside skin"), generally contains the following types of cells:

- Epitheliomuscular cells whose bodies form part of the epithelium but whose bases extend to form muscle fibers in parallel rows. The fibers of the outward-facing cell layer generally run at right angles to the fibers of the inward-facing one. In Anthozoa (anemones, corals, etc.) and Scyphozoa (jellyfish), the mesoglea also contains some muscle cells.
- Cnidocytes, the harpoon-like "nettle cells" that give the phylum Cnidaria its name. These appear between or sometimes on top of the muscle cells.
- Nerve cells. Sensory cells appear between or sometimes on top of the muscle cells, and communicate via synapses (gaps across which chemical signals flow) with motor nerve cells, which lie mostly between the bases of the muscle cells.
- Interstitial cells, which are unspecialized and can replace lost or damaged cells by transforming into the appropriate types. These are found between the bases of muscle cells.

In addition to epitheliomuscular, nerve and interstitial cells, the inward-facing gastroderm ("stomach skin") contains gland cells that secrete digestive enzymes. In some species it also contains low concentrations of cnidocytes, which are used to subdue prey that is still struggling.

The mesoglea contains small numbers of amoeba-like cells, and muscle cells in some species. However, the number of middle-layer cells and types are much lower than in sponges.

Polymorphism

Polymorphism refers to the occurrence of structurally and functionally more than two different types of individuals within the same organism. It is a characteristic feature of Cnidarians, particularly the polyp and medusa forms, or of zooids within colonial organisms like those in Hydrozoa. In Hydrozoans, colonial individuals arising from individuals zooids will take on separate tasks. For example, in *Obelia* there are feeding individuals, the gastrozooids; the individuals capable of asexual reproduction only, the gonozooids, blastostyles and free-living or sexually reproducing individuals, the medusae.

Cnidocytes

These "nettle cells" function as harpoons, since their payloads remain connected to the bodies of the cells by threads. Three types of cnidocytes are known:

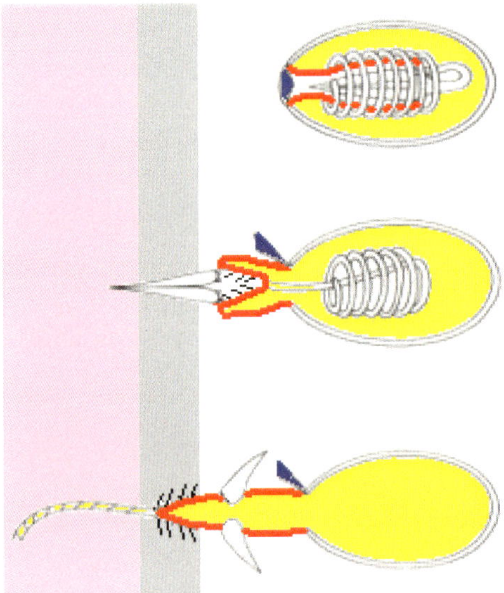

Firing sequence of the cnida in a hydra's nematocyst

Operculum (lid) — Venom

"Finger" that turns inside out — Victim's skin

/// Barbs — Victim's tissues

- Nematocysts inject venom into prey, and usually have barbs to keep them embedded in the victims. Most species have nematocysts.
- Spirocysts do not penetrate the victim or inject venom, but entangle it by means of small sticky hairs on the thread.
- Ptychocysts are not used for prey capture — instead the threads of discharged ptychocysts are used for building protective tubes in which their owners live. Ptychocysts are found only in the order Ceriantharia, tube anemones.

The main components of a cnidocyte are:

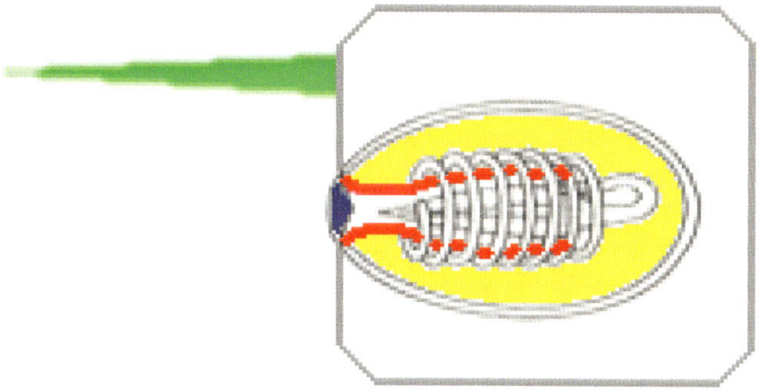

A hydra's nematocyst, before firing. ■ "trigger" cilium

- A cilium (fine hair) which projects above the surface and acts as a trigger. Spirocysts do not have cilia.
- A tough capsule, the cnida, which houses the thread, its payload and a mixture of chemicals that may include venom or adhesives or both.
- A tube-like extension of the wall of the cnida that points into the cnida, like the finger of a rubber glove pushed inwards. When a cnidocyte fires, the finger pops out. If the cell is a venomous nematocyte, the "finger"'s tip reveals a set of barbs that anchor it in the prey.
- The thread, which is an extension of the "finger" and coils round it until the cnidocyte fires. The thread is usually hollow and delivers chemicals from the cnida to the target.
- An operculum (lid) over the end of the cnida. The lid may be a single hinged flap or three flaps arranged like slices of pie.
- The cell body, which produces all the other parts.

It is difficult to study the firing mechanisms of cnidocytes as these structures are small but very complex. At least four hypotheses have been proposed:

- Rapid contraction of fibers round the cnida may increase its internal pressure.
- The thread may be like a coiled spring that extends rapidly when released.
- In the case of *Chironex* (the "sea wasp"), chemical changes in the cnida's contents may cause them to expand rapidly by polymerization.
- Chemical changes in the liquid in the cnida make it a much more concentrated solution, so that osmotic pressure forces water in very rapidly to dilute it. This mechanism has been observed in nematocysts of the class Hydrozoa,

sometimes producing pressures as high as 140 atmospheres, similar to that of scuba air tanks, and fully extending the thread in as little as 2 milliseconds (0.002 second).

Cnidocytes can only fire once, and about 25% of a hydra's nematocysts are lost from its tentacles when capturing a brine shrimp. Used cnidocytes have to be replaced, which takes about 48 hours. To minimise wasteful firing, two types of stimulus are generally required to trigger cnidocytes: nearby sensory cells detect chemicals in the water, and their cilia respond to contact. This combination prevents them from firing at distant or non-living objects. Groups of cnidocytes are usually connected by nerves and, if one fires, the rest of the group requires a weaker minimum stimulus than the cells that fire first.

Locomotion

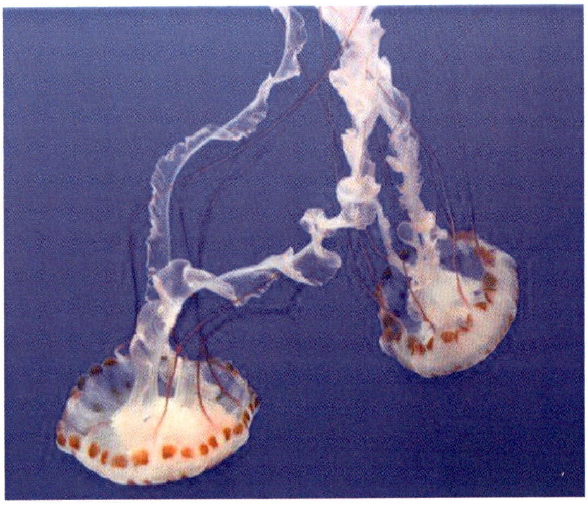

A swimming sea nettle known as the purple-striped jelly (*Chrysaora colorata*)

Medusae swim by a form of jet propulsion: muscles, especially inside the rim of the bell, squeeze water out of the cavity inside the bell, and the springiness of the mesoglea powers the recovery stroke. Since the tissue layers are very thin, they provide too little power to swim against currents and just enough to control movement within currents.

Hydras and some sea anemones can move slowly over rocks and sea or stream beds by various means: creeping like snails, crawling like inchworms, or by somersaulting. A few can swim clumsily by waggling their bases.

Nervous System and Senses

Cnidarians are generally thought to have no brains or even central nervous systems. However, they do have integrative areas of neural tissue that could be considered some form of centralization. Most of their bodies are innervated by decentralized nerve nets that control their swimming musculature and connect with sensory structures, though each clade has slightly different structures. These sensory structures, usually called rhopalia, can generate signals in response to various types of stimuli such as light, pressure, and much more. Medusa usually have several of them around the margin of the bell that work together to control the motor nerve net, that directly innervates the swimming muscles. Most Cnidarians also have a parallel system. In scyphozoans, this takes the form of a diffuse nerve net, which has modulatory effects on the nervous system. As well as forming the "signal cables" between sensory neurons and motoneurons, intermediate neurons in the nerve net can also form ganglia that act as local coordination centers. Communication between nerve cells can occur by chemical synapses or gap junctions in hydrozoans, though gap junctions are not present in all groups. Cnidarians have many of the same neurotransmitters as many animals, including chemicals such as glutamate, GABA, and acetylcholine.

This structure ensures that the musculature is excited rapidly and simultaneously, and can be directly stimulated from any point on the body, and it also is better able to recover after injury.

Medusae and complex swimming colonies such as siphonophores and chondrophores sense tilt and acceleration by means of statocysts, chambers lined with hairs which detect the movements of internal mineral grains called statoliths. If the body tilts in the wrong direction, the animal rights itself by increasing the strength of the swimming movements on the side that is too low. Most species have ocelli ("simple eyes"), which can detect sources of light. However, the agile box jellyfish are unique among Medusae because they possess four kinds of true eyes that have retinas, corneas and lenses. Although the eyes probably do not form images, Cubozoa can clearly distinguish the direction from which light is coming as well as negotiate around solid-colored objects.

Feeding and Excretion

Cnidarians feed in several ways: predation, absorbing dissolved organic chemicals, filtering food particles out of the water, obtaining nutrients from symbiotic algae within their cells, and parasitism. Most obtain the majority of their food from predation but some, including the corals *Hetroxenia* and *Leptogorgia*, depend almost completely on their endosymbionts and on absorbing dissolved nutrients. Cnidaria give their symbiotic algae carbon dioxide, some nutrients, a place in the sun and protection against predators.

Predatory species use their cnidocytes to poison or entangle prey, and those with venomous nematocysts may start digestion by injecting digestive enzymes. The "smell" of fluids from wounded prey makes the tentacles fold inwards and wipe the prey off into the mouth. In medusae the tentacles round the edge of the bell are often short and most of the prey capture is done by "oral arms", which are extensions of the edge of the mouth and are often frilled and sometimes branched to increase their surface area. Medusae often trap prey or suspended food particles by swimming upwards, spreading their tentacles and oral arms and then sinking. In species for which suspended food particles are important, the tentacles and oral arms often have rows of cilia whose beating creates currents that flow towards the mouth, and some produce nets of mucus to trap particles. Their digestion is both intra and extracellular.

Once the food is in the digestive cavity, gland cells in the gastroderm release enzymes that reduce the prey to slurry, usually within a few hours. This circulates through the digestive cavity and, in colonial cnidarians, through the connecting tunnels, so that gastroderm cells can absorb the nutrients. Absorption may take a few hours, and digestion within the cells may take a few days. The circulation of nutrients is driven by water currents produced by cilia in the gastroderm or by muscular movements or both, so that nutrients reach all parts of the digestive cavity. Nutrients reach the outer cell layer by diffusion or, for animals or zooids such as medusae which have thick mesogleas, are transported by mobile cells in the mesoglea.

Indigestible remains of prey are expelled through the mouth. The main waste product of cells' internal processes is ammonia, which is removed by the external and internal water currents.

Respiration

There are no respiratory organs, and both cell layers absorb oxygen from and expel carbon dioxide into the surrounding water. When the water in the digestive cavity becomes stale it must be replaced, and nutrients that have not been absorbed will be expelled with it. Some Anthozoa have ciliated grooves on their tentacles, allowing them to pump water out of and into the digestive cavity without opening the mouth. This improves respiration after feeding and allows these animals, which use the cavity as a hydrostatic skeleton, to control the water pressure in the cavity without expelling undigested food.

Cnidaria that carry photosynthetic symbionts may have the opposite problem, an excess of oxygen, which may prove toxic. The animals produce large quantities of antioxidants to neutralize the excess oxygen.

Regeneration

All cnidarians can regenerate, allowing them to recover from injury and to reproduce asexually. Medusae have limited ability to regenerate, but polyps can do so from small pieces or even collections of separated cells. This enables corals to recover even after apparently being destroyed by predators.

Reproduction

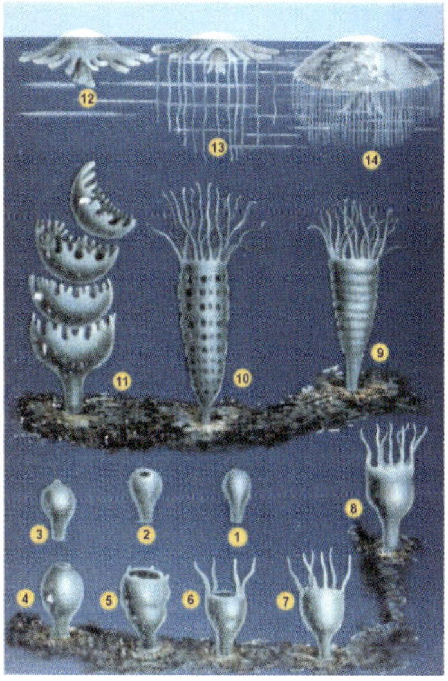

Life cycle of a jellyfish:
1–3 Larva searches for site
4–8 Polyp grows
9–11 Polyp strobilates
12–14 Medusa grows

Sexual

Cnidarian sexual reproduction often involves a complex life cycle with both polyp and medusa stages. For example, in Scyphozoa (jellyfish) and Cubozoa (box jellies) a larva swims until it finds a good site, and then becomes a polyp. This grows normally but then absorbs its tentacles and splits horizontally into a series of disks that become juvenile medusae, a process called strobilation. The juveniles swim off and slowly grow to maturity, while the polyp re-grows and may continue strobilating periodically. The adults have gonads in the gastroderm, and these release ova and sperm into the water in the breeding season.

This phenomenon of succession of differently organized generations (one asexually reproducing, sessile polyp, followed by a free-swimming medusa or a sessile polyp that reproduces sexually) is sometimes called "alternation of asexual and sexual phases" or "metagenesis", but should not be confused with the alternation of generations as found in plants.

Shortened forms of this life cycle are common, for example some oceanic scyphozoans omit the polyp stage completely, and cubozoan polyps produce only one medusa. Hydrozoa have a variety of life cycles. Some have no polyp stages and some (e.g. *hydra*) have no medusae. In some species, the medusae remain attached to the polyp and are responsible for sexual reproduction; in extreme cases these reproductive zooids may not look much like medusae. Meanwhile, life cycle reversal, in which polyps are formed directly from medusae without the involvement of sexual reproduction process, was observed in both Hydrozoa (*Turritopsis dohrnii* and *Laodicea undulata*) and Scyphozoa (*Aurelia* sp.1). Anthozoa have no medusa stage at all and the polyps are responsible for sexual reproduction.

Spawning is generally driven by environmental factors such as changes in the water temperature, and their release is triggered by lighting conditions such as sunrise, sunset or the phase of the moon. Many species of Cnidaria may spawn simultaneously in the same location, so that there are too many ova and sperm for predators to eat more than a tiny percentage — one famous example is the Great Barrier Reef, where at least 110 corals and a few non-cnidarian invertebrates produce enough gametes to turn the water cloudy. These mass spawnings may produce hybrids, some of which can settle and form polyps, but it is not known how long these can survive. In some species the ova release chemicals that attract sperm of the same species.

The fertilized eggs develop into larvae by dividing until there are enough cells to form a hollow sphere (blastula) and then a depression forms at one end (gastrulation) and eventually becomes the digestive cavity. However, in cnidarians the depression forms at the end further from the yolk (at the animal pole), while in bilaterians it forms at the other end (vegetal pole). The larvae, called planulae, swim or crawl by means of cilia. They are cigar-shaped but slightly broader at the "front" end, which is the aboral, vegetal-pole end and eventually attaches to a substrate if the species has a polyp stage.

Anthozoan larvae either have large yolks or are capable of feeding on plankton, and some already have endosymbiotic algae that help to feed them. Since the parents are immobile, these feeding capabilities extend the larvae's range and avoid overcrowding of sites. Scyphozoan and hydrozoan larvae have little yolk and most lack endosymbiotic algae, and therefore have to settle quickly and metamorphose into polyps. Instead, these species rely on their medusae to extend their ranges.

Asexual

All known cnidaria can reproduce asexually by various means, in addition to regenerating after being fragmented. Hydrozoan polyps only bud, while the medusae of some hydrozoans can divide down the middle. Scyphozoan polyps can both bud and split down the middle. In addition to both of these methods, Anthozoa can split horizontally just above the base. Asexual reproduction makes the daughter cnidarian a clone of the adult.

Ctenophora

Ctenophores are fairly simple animals that live only in marine waters; they can be found in most marine habitats, from polar to tropical, inshore to offshore, and from near the surface to the very deep ocean. There are probably about 100-150 species of ctenophores throughout the world's ocean, although most of these are poorly known.

The best-known ctenophores are those that occur near-shore. Such species are typically planktonic, transparent and unpigmented, and most swim by synchronous beating of the eight rows of comb plates. Ctenophores are probably common members of the plankton in most coastal areas worldwide, although they have not been studied in many regions; ctenophores may be seasonally much more abundant in the spring and early summer. Their small size (a few mm to several cm) and transparency make ctenophores relatively inconspicuous, so even though they are not uncommon, in general they are little known. Display of ctenophores in public aquariums throughout the world has been bringing them to the attention of many thousands of viewers in recent years. The simplest tool for collecting ctenophores from a boat or off a dock is a cup-on-a-stick type jellyfish catcher.

Common Features

The Ctenophore phylum has a wide range of body forms, including the flattened, deep-sea platyctenids, in which the adults of most species lack combs, and the coastal beroids, which lack tentacles and prey on other ctenophores by using huge mouths armed with groups of large, stiffened cilia that act as teeth.

Body Layers

Like those of cnidarians, (jellyfish, sea anemones, etc.), ctenophores' bodies consist of a relatively thick, jelly-like mesoglea sandwiched between two epithelia, layers of cells bound by inter-cell connections and by a fibrous basement membrane that they secrete. The epithelia of ctenophores have two layers of cells rather than one, and some of the cells in the upper layer have several cilia per cell.

The outer layer of the epidermis (outer skin) consists of: sensory cells; cells that secrete mucus, which protects the body; and interstitial cells, which can transform into other types of cell. In specialized parts of the body the outer layer also contains colloblasts, found along the surface of tentacles and used in capturing prey, or cells bearing multiple large cilia, for locomotion. The inner layer of the epidermis contains a nerve net, and myoepithelial cells that act as muscles.

The internal cavity forms: a mouth that can usually be closed by muscles; a pharynx ("throat"); a wider area in the center that acts as a stomach; and a system of internal canals. These branch through the mesoglea to the most active parts of the animal: the mouth and pharynx; the roots of the tentacles, if present; all along the underside of each comb row; and four branches round the sensory complex at the far end from the mouth – two of these four branches terminate in anal pores. The inner surface of the cavity is lined with an epithelium, the gastrodermis. The mouth and pharynx have both cilia and well-developed muscles. In other parts of the canal system, the gastrodermis is different on the sides nearest to and furthest from the organ that it supplies. The nearer side is composed of tall nutritive cells that store nutrients in vacuoles (internal compartments), germ cells that produce eggs or sperm, and photocytes that produce bioluminescence. The side furthest from the organ is covered with ciliated cells that circulate water through the canals, punctuated by ciliary rosettes, pores that are surrounded by double whorls of cilia and connect to the mesoglea.

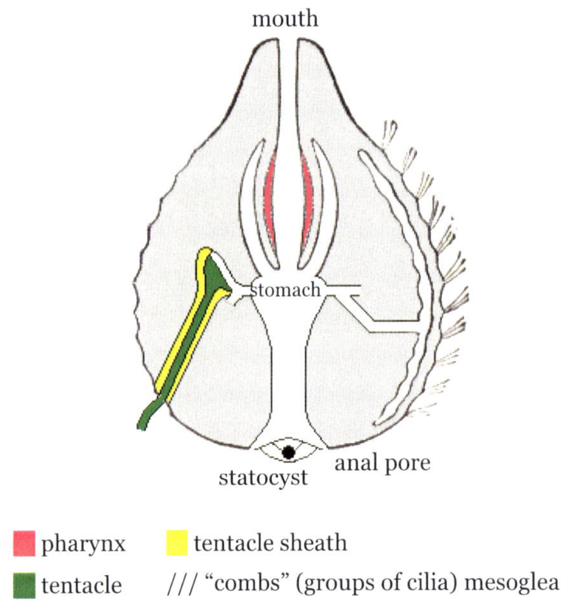

Section through a *Pleurobrachia*-like cydippid ctenophore. Left side shows canal from stomach to tentacle, right shows canal from stomach to comb-row.

Feeding, Excretion and Respiration

When prey is swallowed, it is liquefied in the pharynx by enzymes and by muscular contractions of the pharynx. The resulting slurry is wafted through the canal system by the beating of the cilia, and digested by the nutritive cells. The ciliary rosettes in the canals may help to transport nutrients to muscles in the mesoglea. The anal pores may eject unwanted small particles, but most unwanted matter is regurgitated via the mouth.

Little is known about how ctenophores get rid of waste products produced by the cells. The ciliary rosettes in the gastrodermis may help to remove wastes from the mesoglea, and may also help to adjust the animal's buoyancy by pumping water into or out of the mesoglea.

Locomotion

The outer surface bears usually eight comb rows, called swimming-plates, which are used for swimming. The rows are oriented to run from near the mouth (the "oral pole") to the opposite end (the "aboral pole"), and are spaced more or less evenly around the body, although spacing patterns vary by species and in most species the comb rows extend only part of the distance from the aboral pole towards the mouth. The "combs" (also called "ctenes" or "comb plates") run across each row, and each consists of thousands of unusually long cilia, up to 2 millimeters (0.079 in). Unlike conventional cilia and flagella, which has a filament structure arranged in a 9 + 2 pattern, these cilia are arranged in a 9 + 3 pattern, where the extra compact filament is suspected to have a supporting function. These normally beat so that the propulsion stroke is away from the mouth, although they can also reverse direction. Hence ctenophores usually swim in the direction in which the mouth is eating, unlike jellyfish. When trying to escape predators, one species can accelerate to six times its normal speed; some other species reverse direction as part of their escape behavior, by reversing the power stroke of the comb plate cilia.

It is uncertain how ctenophores control their buoyancy, but experiments have shown that some species rely on osmotic pressure to adapt to water of different densities. Their body fluids are normally as concentrated as seawater. If they enter less dense brackish water, the ciliary rosettes in the body cavity may pump this into the mesoglea to increase its bulk and decrease its density, to avoid sinking. Conversely if they move from brackish to full-strength seawater, the rosettes may pump water out of the mesoglea to reduce its volume and increase its density.

Nervous System and Senses

Ctenophores have no brain or central nervous system, but instead have a nerve net (rather like a cobweb) that forms a ring round the mouth and is densest near structures such as the comb rows, pharynx, tentacles (if present) and the sensory complex furthest from the mouth.

The largest single sensory feature is the aboral organ (at the opposite end from the mouth). Its main component is a statocyst, a balance sensor consisting of a statolith, a tiny grain of calcium carbonate, supported on four bundles of cilia, called "balancers", that sense its orientation. The statocyst is protected by a transparent dome made of long, immobile cilia. A ctenophore does not automatically try to keep the statolith resting equally on all the balancers. Instead its response is determined by the animal's "mood", in other words the overall state of the nervous system. For example, if a ctenophore with trailing tentacles captures prey, it will often put some comb rows into reverse, spinning the mouth towards the prey.

Ctenophore nerve cells and nervous system have a different biochemistry as compared to other animals. For instance they lack the genes and enzymes required to manufacture neurotransmitters like serotonin, dopamine, nitric oxide, acetylcholine, octopamine, noradrenaline and others, which is otherwise seen in all other animals with a nervous system. Also the genes coding for the receptors for each of the neurotransmitters were missing. Therefore, nervous systems may have either been lost in sponges and placozoans, or arisen more than once among metazoans.

Cydippids

Cydippid ctenophores have bodies that are more or less rounded, sometimes nearly spherical and other times more cylindrical or egg-shaped; the common coastal "sea gooseberry", *Pleurobrachia*, sometimes has an egg-shaped body with the mouth at the narrow end, although some individuals are more uniformly round. From opposite sides of the body extends a pair of long, slender tentacles, each housed in a sheath into which it can be withdrawn. Some species of cydippids have bodies that are flattened to various extents, so that they are wider in the plane of the tentacles.

Aulacoctena sp., a cydippid ctenophore

The tentacles of cydippid ctenophores are typically fringed with tentilla ("little tentacles"), although a few genera have simple tentacles without these sidebranches. The tentacles and tentilla are densely covered with microscopic colloblasts that capture prey by sticking to it. Colloblasts are specialized mushroom-shaped cells in the outer layer of the epidermis, and have three main components: a domed head with vesicles (chambers) that contain adhesive; a stalk that anchors the cell in the lower layer of the epidermis or in the mesoglea; and a spiral thread that coils round the stalk and is attached to the head and to the root of the stalk. The function of the spiral thread is uncertain, but it may absorb stress when prey tries to escape, and thus prevent the collobast from being torn apart. In addition to colloblasts, members of the genus *Haeckelia*, which feed mainly on jellyfish, incorporate their victims' stinging nematocytes into their own tentacles – some cnidaria-eating nudibranchs similarly incorporate nematocytes into their bodies for defense. The tentilla of *Euplokamis* differ significantly from those of other cydippids: they contain striated muscle, a cell type otherwise unknown in the phylum Ctenophora; and they are coiled when relaxed, while the tentilla of all other known ctenophores elongate when relaxed. *Euplokamis'* tentilla have three types of movement that are used in capturing prey: they may flick out very quickly (in 40 to 60 milliseconds); they can wriggle, which may lure prey by behaving like small planktonic worms; and they coil round prey. The unique flicking is an uncoiling movement powered by contraction of the striated muscle. The wriggling motion is produced by smooth muscles, but of a highly specialized type. Coiling around prey is accomplished largely by the return of the tentilla to their inactive state, but the coils may be tightened by smooth muscle.

There are eight rows of combs that run from near the mouth to the opposite end, and are spaced evenly round the body. The "combs" beat in a metachronal rhythm rather like that of a Mexican wave. From each balancer in the statocyst a ciliary groove runs out under the dome and then splits to connect with two adjacent comb rows, and in some species runs all the way along the comb rows. This forms a *mechanical* system for transmitting the beat rhythm from the combs to the balancers, via water disturbances created by the cilia.

Lobates

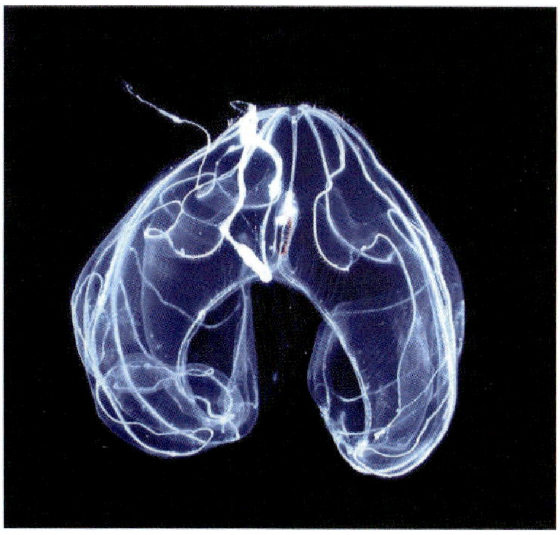

Bathocyroe fosteri a common but fragile deep-sea lobate, oriented mouth down

The Lobata have a pair of lobes, which are muscular, cuplike extensions of the body that project beyond the mouth. Their inconspicuous tentacles originate from the corners of the mouth, running in convoluted grooves and spreading out over the inner surface of the lobes (rather than trailing far behind, as in the Cydippida). Between the lobes on either side of the mouth, many species of lobates have four auricles, gelatinous projections edged with cilia that produce water currents that help direct microscopic prey toward the mouth. This combination of structures enables lobates to feed continuously on suspended planktonic prey.

Lobates have eight comb-rows, originating at the aboral pole and usually not extending beyond the body to the lobes; in species with (four) auricles, the cilia edging the auricles are extensions of cilia in four of the comb rows. Most lobates are quite passive when moving through the water, using the cilia on their comb rows for propulsion, although *Leucothea* has long and active auricles whose movements also contribute to propulsion. Members of the lobate genera *Bathocyroe* and *Ocyropsis* can escape from danger by clapping their lobes, so that the jet of expelled water drives them backwards very quickly. Unlike cydippids, the movements of lobates' combs are coordinated by nerves rather than by water disturbances created by the cilia, yet combs on the same row beat in the same Mexican wave style as the mechanically coordinated comb rows of cydippids and beroids. This may have enabled lobates to grow larger than cydippids and to have shapes that are less egg-like.

An unusual species first described in 2000, *Lobatolampea tetragona*, has been classified as a lobate, although the lobes are "primitive" and the body is medusa-like when floating and disk-like when resting on the sea-bed.

Beroids

Beroe sp. swimming with open mouth, at left. This animal is 3–6 cm long.

The Beroida, also known as Nuda, have no feeding appendages, but their large pharynx, just inside the large mouth and filling most of the saclike body, bears "macrocilia" at the oral end. These fused bundles of several thousand large cilia are able to "bite" off pieces of prey that are too large to swallow whole – almost always other ctenophores. In front of the field of macrocilia, on the mouth "lips" in some species of *Beroe*, is a pair of narrow strips of adhesive epithelial cells on the stomach wall that "zip" the mouth shut when the animal is not feeding, by forming intercellular connections with the opposite adhesive strip. This tight closure streamlines the front of the animal when it is pursuing prey.

Other Body Forms

The Ganeshida have a pair of small oral lobes and a pair of tentacles. The body is circular rather than oval in cross-section, and the pharynx extends over the inner surfaces of the lobes.

The Thalassocalycida, only discovered in 1978 and known from only one species, are medusa-like, with bodies that are shortened in the oral-aboral direction, and short comb-rows on the surface furthest from the mouth, originating from near the aboral pole. They capture prey by movements of the bell and possibly by using two short tentacles.

The Cestida ("belt animals") are ribbon-shaped planktonic animals, with the mouth and aboral organ aligned in the middle of opposite edges of the ribbon. There is a pair of comb-rows along each aboral edge, and tentilla emerging from a groove all along the oral edge, which stream back across most of the wing-like body surface. Cestids can swim by undulating their bodies as well as by the beating of their comb-rows. There are two known species, with worldwide distribution in warm, and warm-temperate waters: *Cestum veneris* ("Venus' girdle") is among the largest ctenophores – up to 1.5 meters (4.9 ft) long, and can undulate slowly or quite rapidly. *Velamen parallelum*, which is typically less than 20 centimeters (0.66 ft) long, can move much faster in what has been described as a "darting motion".

Most Platyctenida have oval bodies that are flattened in the oral-aboral direction, with a pair of tentilla-bearing tentacles on the aboral surface. They cling to and creep on surfaces by everting the pharynx and using it as a muscular "foot". All but one of the known platyctenid species lack comb-rows. Platyctenids are usually cryptically colored, live on rocks, algae, or the body surfaces of other invertebrates, and are often revealed by their long tentacles with many sidebranches, seen streaming off the back of the ctenophore into the current.

Reproduction and Development

Adults of most species can regenerate tissues that are damaged or removed, although only platyctenids reproduce by cloning, splitting off from the edges of their flat bodies fragments that develop into new individuals.

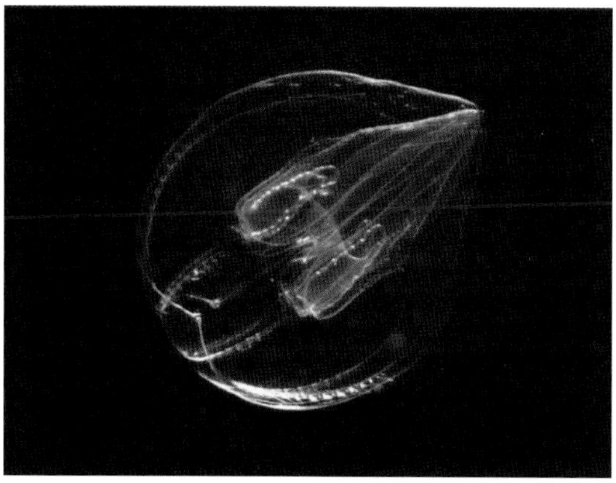

Cydippid larva of *Bolinopsis* sp., a few mm long.

Almost all species are hermaphrodites. In other words, each individual can function as both male and female. Some are simultaneous hermaphrodites, which can produce both eggs and sperm at the same time, while others are sequential hermaphrodites, in which the eggs and sperm mature at different times. The gonads are located in the parts of the internal canal network under the comb rows, and eggs and sperm are released via pores in the epidermis. Fertilization is generally external, but platyctenids use internal fertilization and keep the eggs in brood chambers until they hatch. Self-fertilization has occasionally been seen in species of the genus *Mnemiopsis*, and it is thought that most of the hermaphroditic species are self-fertile.

Development of the fertilized eggs is direct; in other words, there is no distinctive larval form. Juveniles of all groups are generally planktonic, and most species resemble miniature adult cydippids, gradually developing their adult body forms as they grow. In the genus *Beroe*, however, the juveniles have large mouths and, like the adults, lack both tentacles and tentacle sheaths. In some groups, such as the flat, bottom-dwelling platyctenids, the juveniles behave more like true larvae. They live among the plankton and thus occupy a different ecological niche from their parents, only attaining the adult form by a more radical metamorphosis after dropping to the sea-floor.

In contrast to the general hermaphroditism, in two species of the genus *Ocryopsis* individuals remain the same single sex throughout their lives.

At least in some species, juvenile ctenophores appear capable of producing small quantities of eggs and sperm while they are well below adult size, and adults produce eggs and sperm for as long as they have sufficient food. If they run short of food, they first stop producing eggs and sperm, and then shrink in size. When the food supply improves, they grow back to normal size and then resume reproduction. These features make ctenophores capable of increasing their populations very quickly.

Colors and Bioluminescence

Light diffracting along the comb rows of a *Mertensia ovum*. The left tentacle is deployed, the right one retracted.

Undescribed deep-sea species known as "Tortugas red", with trailing tentacles and clearly visible sidebranches, or tentilla.

Most ctenophores that live near the surface are mostly colorless and almost transparent. However some deeper-living species are strongly pigmented, for example the species known as "Tortugas red", which has not yet been formally described. Platyctenids generally live attached to other sea-bottom organisms, and often have similar colors to these host organisms. The gut of the deep-sea genus *Bathocyroe* is red, which hides the bioluminescence of copepods it has swallowed.

The comb rows of most planktonic ctenophores produce a rainbow effect, which is not caused by bioluminescence but by the scattering of light as the combs move. Most species are also bioluminescent, but the light is usually blue or green and can only be seen in darkness. However some significant groups, including all known platyctenids and the cydippid genus *Pleurobrachia*, are incapable of bioluminescence.

When some species, including *Bathyctena chuni*, *Euplokamis stationis* and *Eurhamphaea vexilligera*, are disturbed, they produce secretions (ink) that luminesce at much the same wavelengths as their bodies. Juveniles will luminesce more brightly in relation to their body size than adults, whose luminescence is diffused over their bodies. Detailed statistical investigation has not suggested the function of ctenophores' bioluminescence nor produced any correlation between its exact color and any aspect of the animals' environments, such as depth or whether they live in coastal or mid-ocean waters.

In ctenophores, bioluminescence is caused by the activation of calcium-activated proteins named photoproteins in cells called photocytes, which are often confined to the meridional canals that underlie the eight comb rows. In the genome of *Mnemiopsis leidyi* ten genes encode photoproteins. These genes are co-expressed with opsin genes in the developing photocytes of *Mnemiopsis leidyi*, raising the possibility that light production and light detection may be working together in these animals.

Arthropods

Arthropod, any member of the phylum Arthropoda, the largest phylum in the animal kingdom, which includes such familiar forms as lobsters, crabs, spiders, mites, insects, centipedes, and millipedes. About 84 percent of all known species of animals are members of this phylum. Arthropods are represented in every habitat on Earth and show a great variety of adaptations. Several types live in aquatic environments, and others reside in terrestrial ones; some groups are even adapted for flight.

The distinguishing feature of arthropods is the presence of a jointed skeletal covering composed of chitin (a complex sugar) bound to protein. This nonliving exoskeleton is secreted by the underlying epidermis (which corresponds to the skin of other animals). Arthropods lack locomotory cilia, even in the larval stages, probably because of the presence of the exoskeleton. The body is usually segmented, and the segments bear paired, jointed appendages, from which the name arthropod ("jointed feet") is derived. About one million arthropod species have been described, of which most are insects. This number, however, may be only a fraction of the total. Based on the number of undescribed species collected from the treetops of tropical forests, zoologists have estimated the total number of insect species alone to be as high as 10,000,000. The 30,000 described species of mites may also represent only a fraction of the existing number.

The phylum Arthropoda is commonly divided into four subphyla of extant forms: Chelicerata, Crustacea, Hexapoda, and Myriapoda. Some zoologists believe that arthropods possessing only single-branched appendages, particularly the insects, centipedes, and millipedes, evolved from a separate ancestor and therefore group them within a separate phylum—the Uniramia, or Atelocerata; however, in this treatment these forms are dispersed among several subphyla. In addition, the phylum Arthropoda contains the extinct subphylum Trilobitomorpha. This group is made up of the trilobites, the dominant arthropods in the early Paleozoic seas (542 million to 251 million years ago). Trilobites became extinct during the Permian Period (299 million to 251 million years ago) at the end of the Paleozoic Era.

The myriapods (centipedes, symphylans, millipedes, and pauropods) live beneath stones and logs and in leaf mold; insects are found in all types of terrestrial habitats and some have invaded fresh water. The sea has remained the domain of the crustaceans, however, and only at its very edges are insects (subphylum Hexapoda) found.

The subphylum Crustacea contains mostly marine arthropods, though many of its members, such as the crayfish, have invaded fresh water, and one group, the pill bugs (sow bugs), has become terrestrial, living beneath stones and logs and in leaf mold. In the sea, large crustaceans such as crabs and shrimps are common bottom-dwelling arthropods. Many minute species of crustaceans (particularly the copepods) are an important component of the zooplankton (floating or weakly swimming animals) and serve as food for other invertebrates, fishes, and even whales.

Most members of the subphylum Chelicerata belong to the class Arachnida, containing the spiders, scorpions, ticks, and mites. They are largely terrestrial arthropods, living beneath stones and logs, in leaf mold, and in vegetation, but there are some aquatic mites that live in fresh water and in the sea. There are also many parasitic mites. Two small classes of chelicerates, the Merostomata, containing the horseshoe crabs, and the Pycnogonida, containing the sea spiders, are entirely marine. The merostomes are an ancient group and probably gave rise to the arachnids. Indeed, the earliest known fossil scorpions were aquatic.

Description

Arthropods are invertebrates with segmented bodies and jointed limbs. The exoskeleton or cuticles consists of chitin, a polymer of glucosamine. The cuticle of many crustaceans, beetle mites, and millipedes (except for bristly millipedes) is also biomineralized with calcium carbonate. Calcification of the endosternite, an internal structure used for muscle attachments, also occur in some opiliones.

Diversity

Estimates of the number of arthropod species vary between 1,170,000 and 5 to 10 million and account for over 80 per cent of all known living animal species. The number of species remains difficult to determine. This is due to the census modeling assumptions projected onto other regions in order to scale up from counts at specific locations applied to the whole world. A study in 1992 estimated that there were 500,000 species of animals and plants in Costa Rica alone, of which 365,000 were arthropods.

They are important members of marine, freshwater, land and air ecosystems, and are one of only two major animal groups that have adapted to life in dry environments; the other is amniotes, whose living members are reptiles, birds and mammals. One arthropod sub-group, insects, is the most species-rich member of all ecological guilds in land and freshwater environments. The lightest insects weigh less than 25 micrograms (millionths of a gram), while the heaviest weigh over 70 grams (2.5 oz). Some living crustaceans are much larger; for example, the legs of the Japanese spider crab may span up to 4 metres (13 ft), with the heaviest of all living arthropods being the American lobster, topping out at over 20 kg (44 lbs).

Segmentation

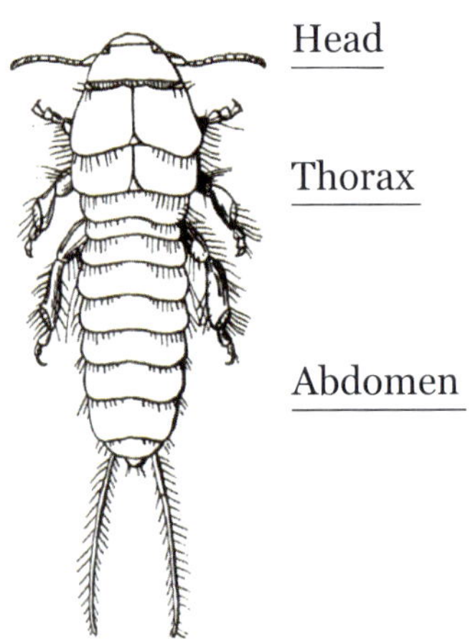

Segments and tagmata of an arthropod

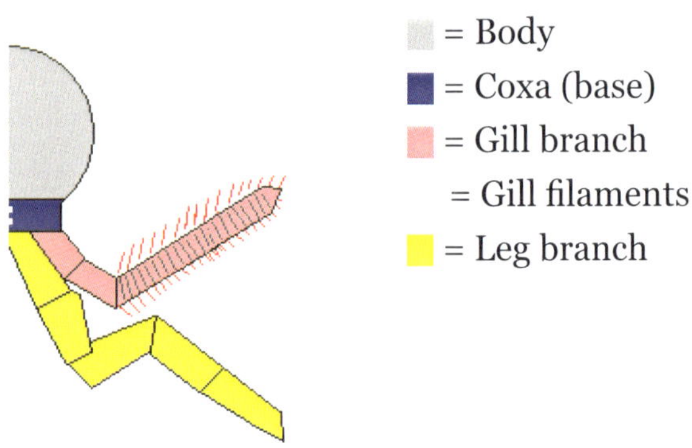

Structure of a biramous appendage

The embryos of all arthropods are segmented, built from a series of repeated modules. The last common ancestor of living arthropods probably consisted of a series of undifferentiated segments, each with a pair of appendages that functioned as limbs. However, all known living and fossil arthropods have grouped segments into tagmata in which segments and their limbs are specialized in various ways.

The three-part appearance of many insect bodies and the two-part appearance of spiders is a result of this grouping; in fact there are no external signs of segmentation in mites. Arthropods also have two body elements that are not part of this serially repeated pattern of segments, an acron at the front, ahead of the mouth, and a telson at the rear, behind the anus. The eyes are mounted on the acron.

Originally it seems that each appendage-bearing segment had two separate pairs of appendages: an upper and a lower pair. These would later fuse into a single pair of biramous appendages, with the upper branch acting as a gill while the lower branch was used for locomotion. In some segments of all known arthropods the appendages have been modified, for example to form gills, mouth-parts, antennae for collecting information, or claws for grasping; arthropods are "like Swiss Army knives, each equipped with a unique set of specialized tools." In many arthropods, appendages have vanished from some regions of the body; it is particularly common for abdominal appendages to have disappeared or be highly modified.

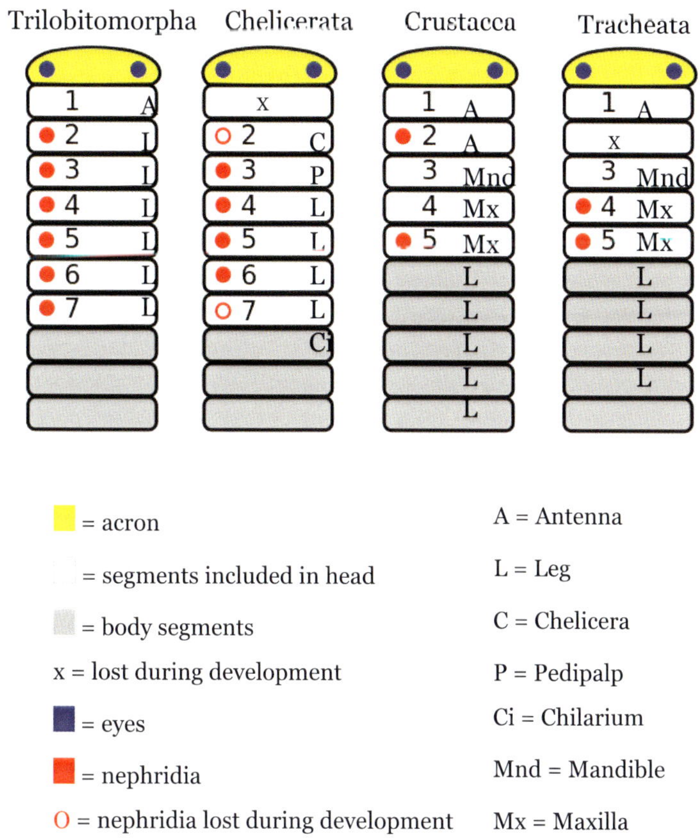

The arthropod head problem

The most conspicuous specialization of segments is in the head. The four major groups of arthropods – Chelicerata (includes spiders and scorpions), Crustacea (shrimps, lobsters, crabs, etc.), Tracheata (arthropods that breathe via channels into their bodies; includes insects and myriapods), and the extinct trilobites – have heads formed of various combinations of segments, with appendages that are missing or specialized in different ways. In addition, some extinct arthropods, such as *Marrella*, belong to none of these groups, as their heads are formed by their own particular combinations of segments and specialized appendages.

Working out the evolutionary stages by which all these different combinations could have appeared is so difficult that it has long been known as "the arthropod head problem".

Exoskeleton

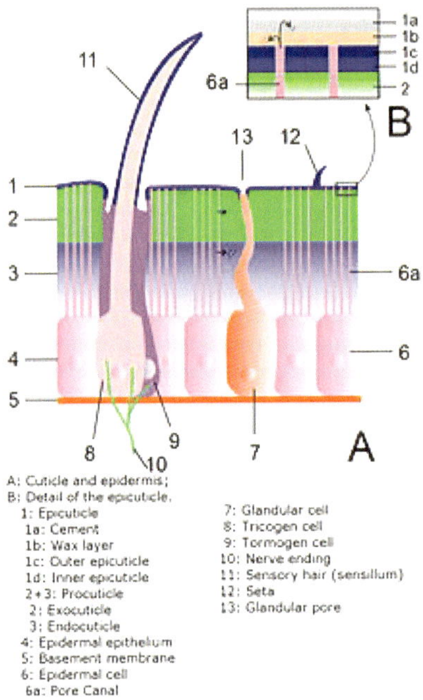

Illustration of an idealized arthropod exoskeleton.

Arthropod exoskeletons are made of cuticle, a non-cellular material secreted by the epidermis. Their cuticles vary in the details of their structure, but generally consist of three main layers: the epicuticle, a thin outer waxy coat that moisture-proofs the other layers and gives them some protection; the exocuticle, which consists of chitin and chemically hardened proteins; and the endocuticle, which consists of chitin and unhardened proteins. The exocuticle and endocuticle together are known as the procuticle. Each body segment and limb section is encased in hardened cuticle. The joints between body segments and between limb sections are covered by flexible cuticle.

The exoskeletons of most aquatic crustaceans are biomineralized with calcium carbonate extracted from the water. Some terrestrial crustaceans have developed means of storing the mineral, since on land they cannot rely on a steady supply of dissolved calcium carbonate. Biomineralization generally affects the exocuticle and the outer part of the endocuticle. Two recent hypotheses about the evolution of biomineralization in arthropods and other groups of animals propose that it provides tougher defensive armor, and that it allows animals to grow larger and stronger by providing more rigid skeletons; and in either case a mineral-organic composite exoskeleton is cheaper to build than an all-organic one of comparable strength.

The cuticle may have setae (bristles) growing from special cells in the epidermis. Setae are as varied in form and function as appendages. For example, they are often used as sensors to detect air or water currents, or contact with objects; aquatic arthropods use feather-like setae to increase the surface area of swimming appendages and to filter food particles out of water; aquatic insects, which are air-breathers, use thick felt-like coats of setae to trap air, extending the time they can spend under water; heavy, rigid setae serve as defensive spines.

Although all arthropods use muscles attached to the inside of the exoskeleton to flex their limbs, some still use hydraulic pressure to extend them, a system inherited from their pre-arthropod ancestors; for example, all spiders extend their legs hydraulically and can generate pressures up to eight times their resting level.

Moulting

Cicada climbing out of its exoskeleton while attached to tree

The exoskeleton cannot stretch and thus restricts growth. Arthropods therefore replace their exoskeletons by moulting, or shedding the old exoskeleton after growing a new one that is not yet hardened. Moulting cycles run nearly continuously until an arthropod reaches full size.

In the initial phase of moulting, the animal stops feeding and its epidermis releases moulting fluid, a mixture of enzymes that digests the endocuticle and thus detaches the old cuticle. This phase begins when the epidermis has secreted a new epicuticle to protect it from the enzymes, and the epidermis secretes the new exocuticle while the old cuticle is detaching. When this stage is complete, the animal makes its body swell by taking in a large quantity of water or air, and this makes the old cuticle split along predefined weaknesses where the old exocuticle was thinnest. It commonly takes several minutes for the animal to struggle out of the old cuticle. At this point, the new one is wrinkled and so soft that the animal cannot support itself and finds it very difficult to move, and the new endocuticle has not yet formed. The animal continues to pump itself up to stretch the new cuticle as much as possible, then hardens the new exocuticle and eliminates the excess air or water. By the end of this phase, the new endocuticle has formed. Many arthropods then eat the discarded cuticle to reclaim its materials.

Because arthropods are unprotected and nearly immobilized until the new cuticle has hardened, they are in danger both of being trapped in the old cuticle and of being attacked by predators. Moulting may be responsible for 80 to 90% of all arthropod deaths.

Internal Organs

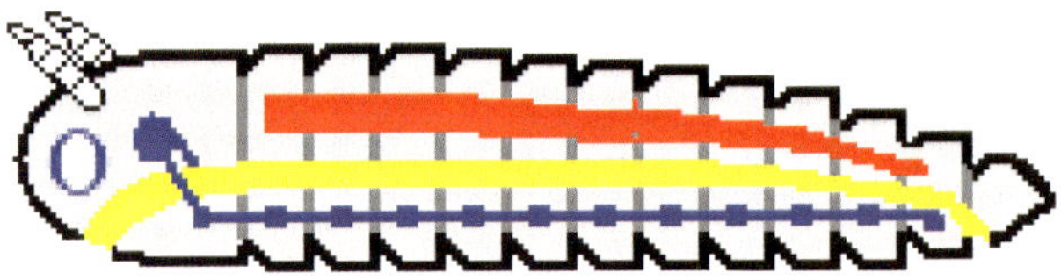

■ = heart
■ = gut
■ = brain / ganglia
o = eye

Basic arthropod body structure

Arthropod bodies are also segmented internally, and the nervous, muscular, circulatory, and excretory systems have repeated components. Arthropods come from a lineage of animals that have a coelom, a membrane-lined cavity between the gut and the body wall that accommodates the internal organs. The strong, segmented limbs of arthropods eliminate the need for one of the coelom's main ancestral functions, as a hydrostatic skeleton, which muscles compress in order to change the animal's shape and thus enable it to move. Hence the coelom of the arthropod is reduced to small areas around the reproductive and excretory systems. Its place is largely taken by a hemocoel, a cavity that runs most of the length of the body and through which blood flows.

Arthropods have open circulatory systems, although most have a few short, open-ended arteries. In chelicerates and crustaceans, the blood carries oxygen to the tissues, while hexapods use a separate system of tracheae. Many crustaceans, but few chelicerates and tracheates, use respiratory pigments to assist oxygen transport. The most common respiratory pigment in arthropods is copper-based hemocyanin; this is used by many crustaceans and a few centipedes. A few crustaceans and insects use iron-based hemoglobin, the respiratory pigment used by vertebrates. As with other invertebrates, the respiratory pigments of those arthropods that have them are generally dissolved in the blood and rarely enclosed in corpuscles as they are in vertebrates.

The heart is typically a muscular tube that runs just under the back and for most of the length of the hemocoel. It contracts in ripples that run from rear to front, pushing blood forwards. Sections not being squeezed by the heart muscle are expanded either by elastic ligaments or by small muscles, in either case connecting the heart to the body wall. Along the heart run a series of paired ostia, non-return valves that allow blood to enter the heart but prevent it from leaving before it reaches the front.

Arthropods have a wide variety of respiratory systems. Small species often do not have any, since their high ratio of surface area to volume enables simple diffusion through the body surface to supply enough oxygen. Crustacea usually have gills that are modified appendages. Many arachnids have book lungs. Tracheae, systems of branching tunnels that run from the openings in the body walls, deliver oxygen directly to individual cells in many insects, myriapods and arachnids.

Living arthropods have paired main nerve cords running along their bodies below the gut, and in each segment the cords form a pair of ganglia from which sensory and motor nerves run to other parts of the segment. Although the pairs of ganglia in each segment often appear physically fused, they are connected by commissures (relatively large bundles of nerves), which give arthropod nervous systems a characteristic “ladder-like” appearance. The brain is in the head, encircling and mainly *above* the esophagus. It consists of the fused ganglia of the acron and one or two of the foremost segments that form the head – a total of three pairs of ganglia in most arthropods, but only two in chelicerates, which do not have antennae or the ganglion connected to them. The ganglia of other head segments are often close to the brain and function as part of it. In insects these other head ganglia combine into a pair of subesophageal ganglia, under and behind the esophagus. Spiders take this process a step further, as *all* the segmental ganglia are incorporated into the subesophageal ganglia, which occupy most of the space in the cephalothorax (front “super-segment”).

There are two different types of arthropod excretory systems. In aquatic arthropods, the end-product of biochemical reactions that metabolise nitrogen is ammonia, which is so toxic that it needs to be diluted as much as possible with water. The ammonia is then eliminated via any permeable membrane, mainly through the gills. All crustaceans use this system, and its high consumption of water may be responsible for the relative lack of success of crustaceans as land animals. Various groups of terrestrial arthropods have independently developed a different system: the end-product of nitrogen metabolism is uric acid, which can be excreted as dry material; the Malpighian tubule system filters the uric acid and other nitrogenous waste out of the blood in the hemocoel, and dumps these materials into the hindgut, from which they are expelled as feces. Most aquatic arthropods and some terrestrial ones also have organs called nephridia (“little kidneys”), which extract other wastes for excretion as urine.

Senses

Optical

The stiff cuticles of arthropods would block out information about the outside world, except that they are penetrated by many sensors or connections from sensors to the nervous system. In fact, arthropods have modified their cuticles into elaborate arrays of sensors. Various touch sensors, mostly setae, respond to different levels of force, from strong contact to very weak air currents. Chemical sensors provide equivalents of taste and smell, often by means of setae. Pressure sensors often take the form of membranes that function as eardrums, but are connected directly to nerves rather than to auditory ossicles. The antennae of most hexapods include sensor packages that monitor humidity, moisture and temperature.

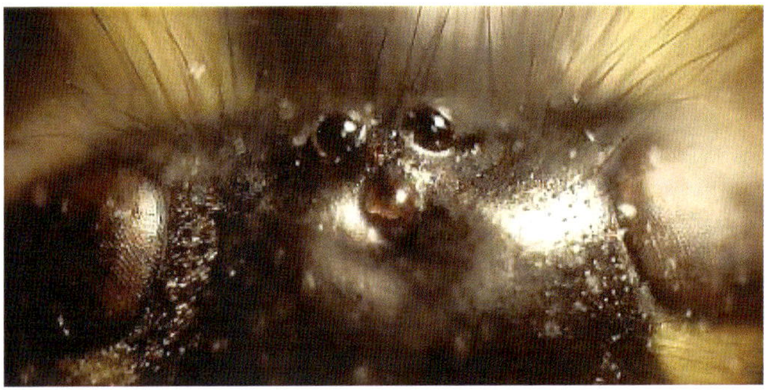

Head of a wasp with three ocelli (centre), and compound eyes at the left and right

Most arthropods have sophisticated visual systems that include one or more usually both of compound eyes and pigment-cup ocelli ("little eyes"). In most cases ocelli are only capable of detecting the direction from which light is coming, using the shadow cast by the walls of the cup. However, the main eyes of spiders are pigment-cup ocelli that are capable of forming images, and those of jumping spiders can rotate to track prey.

Compound eyes consist of fifteen to several thousand independent ommatidia, columns that are usually hexagonal in cross section. Each ommatidium is an independent sensor, with its own light-sensitive cells and often with its own lens and cornea. Compound eyes have a wide field of view, and can detect fast movement and, in some cases, the polarization of light. On the other hand, the relatively large size of ommatidia makes the images rather coarse, and compound eyes are shorter-sighted than those of birds and mammals – although this is not a severe disadvantage, as objects and events within 20 centimetres (7.9 in) are most important to most arthropods. Several arthropods have color vision, and that of some insects has been studied in detail; for example, the ommatidia of bees contain receptors for both green and ultra-violet.

Most arthropods lack balance and acceleration sensors, and rely on their eyes to tell them which way is up. The self-righting behavior of cockroaches is triggered when pressure sensors on the underside of the feet report no pressure. However, many malacostracan crustaceans have statocysts, which provide the same sort of information as the balance and motion sensors of the vertebrate inner ear.

The proprioceptors of arthropods, sensors that report the force exerted by muscles and the degree of bending in the body and joints, are well understood. However, little is known about what other internal sensors arthropods may have.

Olfaction

Reproduction and Development

Compsobuthus werneri female with young (white)

A few arthropods, such as barnacles, are hermaphroditic, that is, each can have the organs of both sexes. However, individuals of most species remain of one sex their entire lives. A few species of insects and crustaceans can reproduce by parthenogenesis, especially if conditions favor a "population explosion". However, most arthropods rely on sexual reproduction, and parthenogenetic species often revert to sexual reproduction when conditions become less favorable. Aquatic arthropods may breed by external fertilization, as for example frogs do, or by internal fertilization, where the ova remain in the female's body and the sperm must somehow be inserted. All known terrestrial arthropods use internal fertilization. Opiliones (harvestmen), millipedes, and some crustaceans use modified appendages such as gonopods or penises to transfer the sperm directly to the female. However, most male terrestrial arthropods produce spermatophores, waterproof packets of sperm, which the females take into their bodies. A few such species rely on females to find spermatophores that have already been deposited on the ground, but in most cases males only deposit spermatophores when complex courtship rituals look likely to be successful.

Most arthropods lay eggs, but scorpions are ovoviparous: they produce live young after the eggs have hatched inside the mother, and are noted for prolonged maternal care. Newly born arthropods have diverse forms, and insects alone cover the range of extremes. Some hatch as apparently miniature adults (direct development), and in some cases, such as silverfish, the hatchlings do not feed and may be helpless until after their first moult. Many insects hatch as grubs or caterpillars, which do not have segmented limbs or hardened cuticles, and metamorphose into adult forms by entering an inactive phase in which the larval tissues are broken down and re-used to build the adult body. Dragonfly larvae have the typical cuticles and jointed limbs of arthropods but are flightless water-breathers with extendable jaws. Crustaceans commonly hatch as tiny nauplius larvae that have only three segments and pairs of appendages.

The nauplius larva of a penaeid shrimp

Echinoderms

Echinoderm, any of a variety of invertebrate marine animals belonging to the phylum Echinodermata, characterized by a hard, spiny covering or skin. Beginning with the dawn of the Cambrian Period (542 million to 488 million years ago), echinoderms have a rich fossil history and are well represented by many bizarre groups, most of which are now extinct. Living representatives include the classes Crinoidea (sea lilies and feather stars), Echinoidea (sea urchins), Holothuroidea (sea cucumbers), Asteroidea (starfishes, or sea stars), Ophiuroidea (basket stars and serpent stars, or brittle stars), and the recently discovered Concentricycloidea (sea daisies).

Echinoderms have been recognized since ancient times; echinoids, for example, were used extensively by Greeks and Romans for medicinal purposes and as food. During the Middle Ages, fossil echinoids and parts of fossil crinoids were objects of superstition. In the early part of the 19th century, Echinodermata was recognized as a distinct group of animals and was occasionally associated with the cnidarians and selected other phyla in a division of the animal kingdom known as the Radiata; the concept of a superphylum called Radiata is no longer valid.

Echinoderms are separated into 21 classes, based mainly on differences in skeletal structures. The number of extant species exceeds 6,500, and approximately 13,000 fossil species have been described.

Anatomy and Physiology

Echinoderms evolved from animals with bilateral symmetry. Although adult echinoderms possess pentaradial, or five-sided, symmetry, echinoderm larvae are ciliated, free-swimming organisms that organize in bilateral symmetry which makes them look like embryonic chordates. Later, the left side of the body grows at the expense of the right side, which is eventually absorbed. The left side then grows in a pentaradially symmetric fashion, in which the body is arranged in five parts around a central axis. Within the Asterozoa, there can be a few exceptions from the rule. The starfish genus Leptasterias normally have six arms, although five armed individuals can occur. Also the Brisingida have six armed species. Amongst the brittle stars, six armed species such as Ophiothela danae, Ophiactis savignyi and Ophionotus hexactis exists, and Ophiacantha vivipara often has more than six.

Echinoderms exhibit secondary radial symmetry in portions of their body at some stage of life. This, however, is an adaptation to their sessile existence. They developed from other members of the Bilateria and exhibit bilateral symmetry in their larval stage. Many crinoids and some seastars exhibit symmetry in multiples of the basic five, with starfish such as *Labidiaster annulatus* known to possess up to fifty arms, and the sea-lily *Comaster schlegelii* having two hundred.

Skin and Skeleton

Echinoderms have a mesodermal skeleton composed of calcareous plates or ossicles. Each one of these, even the articulating spine of a sea urchin, is composed mineralogically of a crystal of calcite. If solid, these would form a heavy skeleton, so they have a sponge-like porous structure known as stereom. Ossicles may be fused together, as in the test of sea urchins, or may articulate with each other as in the arms of sea stars, brittle stars and crinoids. The ossicles may be flat plates or bear external projections in the form of spines, granules or warts and they are supported by a tough epidermis (skin). Skeletal elements are also deployed in some specialized ways, such as the "Aristotle's lantern" mouthparts of sea urchins used for grinding, the supportive stalks of crinoids and the structural "lime ring" of sea cucumbers.

A brittle star, *Ophionereis reticulata*

A sea cucumber from Malaysia

Starfish exhibit a wide range of colours

Strongylocentrotus purpuratus, a well-armoured sea urchin

Despite the robustness of the individual skeletal modules complete skeletons of starfish, brittle stars and crinoids are rare in the fossil record. This is because they quickly disarticulate (disconnect from each other) once the encompassing skin rots away, and in the absence of tissue there is nothing to hold the plates together. The modular construction is a result of the growth system employed by echinoderms, which adds new segments at the centre of the radial limbs, pushing the existing plates outwards and lengthening the arms. Sea urchins on the other hand are often well preserved in chalk beds or limestone. During fossilization, the cavities in the stereom are filled in with calcite that is in crystalline continuity with the surrounding material. On fracturing such rock, distinctive cleavage patterns can be seen and sometimes even the intricate internal and external structure of the test.

Crinoid on a coral reef

The epidermis consists of cells responsible for the support and maintenance of the skeleton, as well as pigment cells, mechanoreceptor cells (which detect motion on the animal's surface), and sometimes gland cells which secrete sticky fluids or even toxins. The varied and often vivid colours of echinoderms are produced by the action of skin pigment cells. These are produced by a variable combination of coloured pigments, such as the dark melanin, red carotinoids, and carotene proteins, which can be blue, green, or violet. These may be light-sensitive, and as a result many echinoderms change appearance completely as night falls. The reaction can happen quickly — the sea urchin *Centrostephanus longispinus* changes from jet black to grey-brown in just fifty minutes when exposed to light.

One characteristic of most echinoderms is a special kind of tissue known as "catch connective tissue". This collagenous material can change its mechanical properties in a few seconds or minutes through nervous control rather than by muscular means. This tissue enables a starfish to change from moving flexibly around the seabed to becoming rigid while prying open a bivalve mollusc or preventing itself from being extracted from a crevice. Similarly, sea urchins can lock their normally mobile spines rigidly as a defensive mechanism when attacked.

The Water Vascular System

Echinoderms possess a unique water vascular system. This is a network of fluid-filled canals derived from the coelom (body cavity) that function in gas exchange, feeding, sensory reception and locomotion. This system varies between different classes of echinoderm but typically opens to the exterior through a sieve-like madreporite on the aboral (upper) surface of the animal. The madreporite is linked to a slender duct, the stone canal, which extends to a ring canal that encircles the mouth or oesophagus. From this, radial canals extend along the arms of asteroids and adjoin the test in the ambulacral areas of echinoids. Short lateral canals branch off the radial canals, each one ending in an ampulla. Part of the ampulla can protrude through a pore (or a pair of pores in sea urchins) to the exterior and is known as a podium or tube feet. The water vascular system assists with the distribution of nutrients throughout the animal's body and is most obviously expressed in the tube feet which can be extended or contracted by the redistribution of fluid between the foot and the internal sac.

The organization of the system is somewhat different in ophiuroids where the medreporite may be on the oral surface and the podia lack suckers. In holothuroids, the podia may be reduced or absent and the madreporite opens into the body cavity so that the circulating liquid is coelomic fluid rather than sea water. The arrangements in crinoids is similar to asteroids but the tube feet lack suckers and are used to pass food particles captured by the arms towards the central mouth. In the asteroids, the same wafting motion is employed to move the animal across the ground. Sea urchins use their feet to prevent the larvae of encrusting organisms from settling on their surfaces; potential settlers are moved to the urchin's mouth and eaten. Some burrowing sea stars extend their elongated dorsal tube feet to the surface of the sand or mud above and use them to absorb oxygen from the water column.

Other Organs

Echinoderms possess a simple digestive system which varies according to the animal's diet. Starfish are mostly carnivorous and have a mouth, oesophagus, two-part stomach, intestine and rectum, with the anus located in the centre of the aboral body surface. With a few exceptions, the members of the order Paxillosida do not possess an anus. In many species of starfish, the large cardiac stomach can be everted and digest food outside the body. In other species, whole food items such as molluscs may be ingested. Brittle stars have a blind gut with no intestine or anus. They have varying diets and expel food waste through their mouth. Sea urchins are herbivores and use their specialised mouthparts to graze, tear and chew algae and sometimes other animal or vegetable material. They have an oesophagus, a large stomach and a rectum with the anus at the apex of the test. Sea cucumbers are mostly detritivores, sorting through the sediment with their buccal tentacles which are modified tube feet. Sand and mud accompanies their food through their simple gut which has a long coiled intestine and a capacious cloaca. Crinoids are passive suspension feeders, catching plankton with their outstretched arms. Boluses of mucus-trapped food are passed to the mouth which is linked to the anus by a loop consisting of a short oesophagus and longer intestine.

The coelomic cavities of echinoderms are complex. Aside from the water vascular system, echinoderms have a haemal coelom (or haemal system, the "haemal" being a misnomer), a perivisceral coelom, a gonadal coelom and often also a perihaemal coelom (or perihaemal system). During development, echinoderm coelom is divided in metacoel, mesocoel and protocoel (also called somatocoel, hydrocoel and axocoel, respectively). The water vascular system, haemal system and perihaemal system form the tubular coelomic system. Echinoderms are an exception having both a coelomic circulatory system (i.e., the water vascular system) and a haemal circulatory system (i.e., the haemal and perihaemal systems).

Haemal and perihaemal systems are derived from the coelom and form an open and reduced circulatory system. This usually consists of a central ring and five radial vessels. There is no true heart and the blood often lacks any respiratory pigment. Gaseous exchange occurs via dermal branchae or papulae in starfish, genital bursae in brittle stars, peristominal gills in sea urchins and cloacal trees in sea cucumbers. Exchange of gases also takes place through the tube feet. Echinoderms lack specialized excretory (waste disposal) organs and so nitrogenous waste, chiefly in the form of ammonia, diffuses out through the respiratory surfaces.

The coelomic fluid contains the coelomocytes, or immune cells. There are several types of immune cells, which vary among classes and species. All classes possess a type of phagocytic amebocyte, which engulf invading particles and infected cells, aggregate or clot, and may be involved in cytotoxicity. These cells are usually larger and granular, and are suggested to be a main line of defense against potential pathogens. Depending on the class, echinoderms may have spherule cells (for cytotoxicity, inflammation, and anti-bacterial activity), vibratile cells (for coelomic fluid movement and clotting), and crystal cells (potential osmoregulatory cells in sea cucumbers),. The coelomocytes also secrete Anti-Microbial Peptides (AMPs) against bacteria, and have a set of lectins and complement proteins as part of an innate immune system that is still being characterized.

Echinoderms have a simple radial nervous system that consists of a modified nerve net consisting of interconnecting neurons with no central brain, although some do possess ganglia. Nerves radiate from central rings around the mouth into each arm or along the body wall; the branches of these nerves coordinate the movements of the organism and the synchronisation of the tube feet. Starfish have sensory cells in the epithelium and have simple eyespots and touch-sensitive tentacle-like tube feet at the tips of their arms. Sea urchins have no particular sense organs but do have statocysts that assist in gravitational orientation, and they have sensory cells in their epidermis, particularly in the tube feet, spines and pedicellariae. Brittle stars, crinoids and sea cucumbers in general do not have sensory organs but some burrowing sea cucumbers of the order Apodida have a single statocyst adjoining each radial nerve and some have an eyespot at the base of each tentacle.

The gonads occupy much of the body cavities of sea urchins and sea cucumbers, while the less voluminous crinoids, brittle stars and starfish have two gonads in each arm. While the ancestral condition is considered to be the possession of one genital aperture, many organisms have multiple gonopores through which eggs or sperm may be released.

Regeneration

Sunflower star regenerating several arms

Many echinoderms have remarkable powers of regeneration. Many species routinely autotomize and regenerate arms and viscera. Sea cucumbers often discharge parts of their internal organs if they perceive themselves to be threatened. The discharged organs and tissues are regenerated over the course of several months. Sea urchins are constantly replacing spines lost through damage. Sea stars and sea lilies readily lose and regenerate their arms. In most cases, a single severed arm cannot grow into a new starfish in the absence of at least part of the disc. However, in a few species a single arm can survive and develop into a complete individual and in some species, the arms are intentionally detached for the purpose of asexual reproduction. During periods when they have lost their digestive tracts, sea cucumbers live off stored nutrients and absorb dissolved organic matter directly from the water.

The regeneration of lost parts involves both epimorphosis and morphallaxis. In epimorphosis stem cells—either from a reserve pool or those produced by dedifferentiation—form a blastema and generate new tissues. Morphallactic regeneration involves the movement and remodelling of existing tissues to replace lost parts. Direct transdifferentiation of one type of tissue to another during tissue replacement is also observed.

The robust larval growth is responsible for many echinoderms being used as popular model organisms in developmental biology.

Reproduction

Sexual Reproduction

Echinoderms become sexually mature after approximately two to three years, depending on the species and the environmental conditions. They are nearly all gonochoric, though a few species are hermaphroditic. The eggs and sperm cells are typically released into open water, where fertilization takes place. The release of sperm and eggs is synchronised in some species, usually with regard to the lunar cycle. In other species, individuals may aggregate during the reproductive season, thereby increasing the likelihood of successful fertilisation. Internal fertilisation has currently been observed in three species of sea star, three brittle stars and a deep water sea cucumber. Even at abyssal depths, where no light penetrates, synchronisation of reproductive activity in echinoderms is surprisingly frequent.

Some echinoderms brood their eggs. This is especially common in cold water species where planktonic larvae might not be able to find sufficient food. These retained eggs are usually few in number and are supplied with large yolks to nourish the developing embryos. In starfish, the female may carry the eggs in special pouches, under her arms, under her arched body or even in her cardiac stomach. Many brittle stars are hermaphrodites. Egg brooding is quite common and usually takes place in special chambers on their oral surfaces, but sometimes the ovary or coelom is used. In these starfish and brittle stars, direct development without passing through a bilateral larval stage usually takes place. A few sea urchins and one species of sand dollar carry their eggs in cavities, or near their anus, holding them in place with their spines. Some sea cucumbers use their buccal tentacles to transfer their eggs to their underside or back where they are retained. In a very small number of species, the eggs are retained in the coelom where they develop viviparously, later emerging through ruptures in the body wall. In some species of crinoid, the embryos develop in special breeding bags, where the eggs are held until sperm released by a male happens to find them.

Asexual Reproduction

One species of seastar, *Ophidiaster granifer*, reproduces asexually by parthenogenesis. In certain other asterozoans, the adults reproduce asexually for a while before they mature after which time they reproduce sexually. In most of these species, asexual reproduction is by transverse fission with the disc splitting in two. Regrowth of both the lost disc area and the missing arms occur so that an individual may have arms of varying lengths. Though in most species at least part of the disc is needed for complete regeneration, in a few species of sea stars, a single severed arm can grow into a complete individual over a period of several months. In at least some of these species, they actively use this as a method of asexual reproduction. A fracture develops on the lower surface of the arm and the arm pulls itself free from the body which holds onto the substrate during the process. During the period of regrowth, they have a few tiny arms and one large arm, thus often being referred to as "comets".

Asexual reproduction by transverse fission has also been observed in adult sea cucumbers. *Holothuria parvula* uses this method frequently, an individual splitting into two a little in front of the midpoint. The two halves each regenerate their missing organs over a period of several months but the missing genital organs are often very slow to develop.

The larvae of some echinoderm species are capable of asexual reproduction. This has long been known to occur among starfish and brittle stars but has been more recently observed in a sea cucumber, a sand dollar and a sea urchin. These species belong to four of the major classes of echinoderms except crinozoans (as of 2011). Asexual reproduction in the planktonic larvae occurs through numerous modes. They may autotomise parts that develop into secondary larvae, grow buds or undergo paratomy. The parts that are autotomised or the buds may develop directly into fully formed larvae or may develop through a gastrula or even a blastula stage. The parts that develop into the new larvae vary from the preoral hood (a mound like structure above the mouth), the side body wall, the postero-lateral arms or their rear ends.

The process of cloning is a cost borne by the larva both in resources as well as in development time. Larvae have been observed to undergo this process when food is plentiful or temperature conditions are optimal. It has also been suggested that cloning may occur to make use of the tissues that are normally lost during metamorphosis. Recent research has shown that the larvae of some sand dollars clone themselves when they detect predators (by sensing dissolved fish mucus). Asexual reproduction produces many smaller larvae that escape better from planktivorous fish.

Larval Development

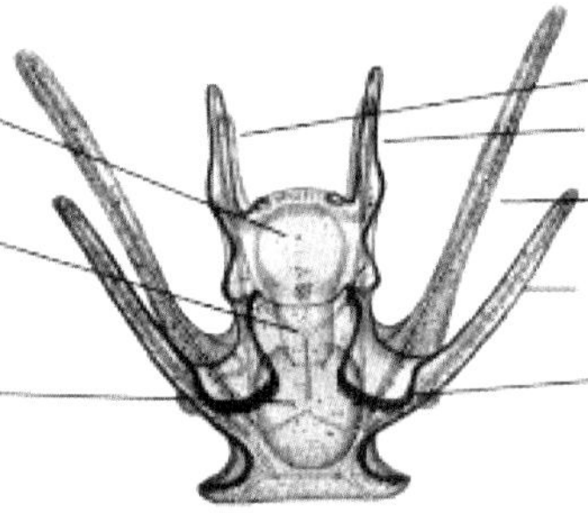

An echinopluteus larva with larval arms

The development of an echinoderm begins with a bilaterally symmetrical embryo, with a coeloblastula developing first. Gastrulation marks the opening of the "second mouth" that places echinoderms within the deuterostomes, and the mesoderm, which will host the skeleton, migrates inwards. The secondary body cavity, the coelom, forms by the partitioning of three body cavities. The larvae are mostly planktonic but in some species the eggs are retained inside the female and in some, the larvae are also brooded by the female.

The larvae of echinoderms pass through a number of stages and these have specific names derived from the taxonomic names of the adults or from their appearance. For example, a sea urchin has an 'echinopluteus' larva while a brittle star has an 'ophiopluteus' larva. A starfish has a 'bipinnaria' larva but this later develops into a multi-armed 'brachiolaria' larva. A sea cucumber larva is an 'auricularia' while a crinoid one is a 'vitellaria'. All these larvae are bilaterally symmetrical and have bands of cilia with which they swim and some, usually known as 'pluteus' larvae, have arms. When fully developed they settle on the seabed to undergo metamorphosis and the larval arms and gut degenerate. The left hand side of the larva develops into the oral surface of the juvenile while the right side becomes the aboral surface. At this stage the bilateral symmetry is lost and radial symmetry develops.

The planktotrophic larva is considered to be the ancestral larval type for echinoderms but after 500 million years of larval evolution, about 68% of species whose development is known have a lecithotrophic larval type. The provision of a yolk-sac means that smaller numbers of eggs are produced, the larvae have a shorter development period, smaller dispersal potential but a greater chance of survival. There seems to be an evolutionary trend towards a "lower-risk–lower-gain" strategy of direct development.

Distribution and Habitat

Echinoderms are globally distributed in almost all depths, latitudes and environments in the ocean. They reach highest diversity in reef environments but are also widespread on shallow shores, around the poles — refugia where crinoids are at their most abundant

— and throughout the deep ocean, where bottom-dwelling and burrowing sea cucumbers are common — sometimes accounting for up to 90% of organisms. While almost all echinoderms are benthic — that is, they live on the sea floor — some sea-lilies can swim at great velocity for brief periods of time, and a few deep-sea sea cucumbers are fully floating. Some crinoids are pseudo-planktonic, attaching themselves to floating logs and debris, although this behaviour was exercised most extensively in the Paleozoic, before competition from such organisms as barnacles restricted the extent of the behaviour.

The larvae of echinoderms, especially starfish and sea urchins, are pelagic, and with the aid of ocean currents can be transported for great distances, reinforcing the global distribution of the phylum.

Annelid

The annelids include earthworms, polychaete worms, and leeches. All members of the group are to some extent segmented, in other words, made up of segments that are formed by subdivisions that partially transect the body cavity. Segmentation is also called metamerism. Segments each contain elements of such body systems as circulatory, nervous, and excretory tracts. Metamerism increases the efficiency of body movement by allowing the effect of muscle contraction to be extremely localized, and it makes possible the development of greater complexity in general body organization.

Besides being segmented, the body wall of annelids is characterized by being made up of both circular and longitudinal muscle fibers surrounded by a moist, acellular cuticle that is secreted by an epidermal epithelium. All annelids except leeches also have chitonous hair-like structures, called setae, projecting from their cuticle. Sometimes the setae are located on paddle-like appendages called parapodia.

Annelids are schizocoelous and with a large and well-developed true coelom (i.e., one that is lined with mesoderm). Except in leeches, the coelom is partially subdivided by septa. Hydrostatic pressure is maintained across segments and helps maintain body rigidity, allowing muscle contractions to bend the body without collapsing it.

The internal organs of annelids are well developed. They include a closed, segmentally-arranged circulatory system. The digestive system is a complete tube with mouth and anus. Gases are exchanged through the skin, or sometimes through specialized gills or modified parapodia. Each segment typically contains a pair of nephridia. The nervous system includes a pair of cephalic ganglia attached to double nerve cords that run the length of the animal along the ventral body wall, with ganglia and branches in each segment. Annelids have some combination of tactile organs, chemoreceptors, balance receptors, and photoreceptors; some forms have fairly well developed eyes, including lenses.

Annelids may be monoecious or dioecious. Larva may or may not be present; if present they are of the trochophore type. Some forms also reproduce asexually. They are protostomes, with spiral cleavage.

Members of the Phylum Annelida can be found throughout the world, in marine, freshwater, and terrestrial environments. Ecologically, they range from passive filter feeders to voracious and active predators.

Description

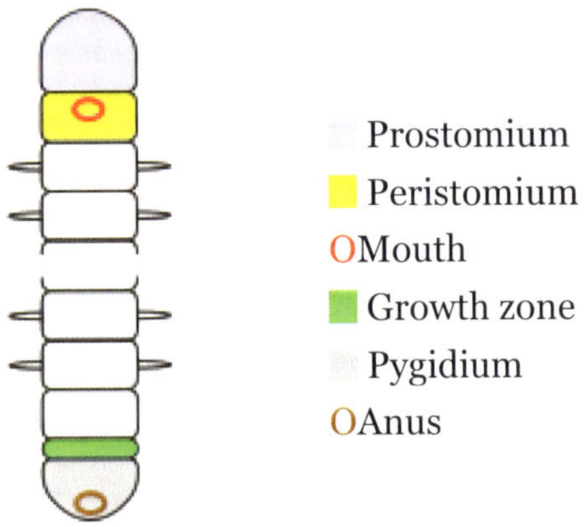

Segments of an annelid

Most of an annelid's body consists of segments that are practically identical, having the same sets of internal organs and external chaetae and, in some species, appendages. However, the frontmost and rearmost sections are not regarded as true segments as they do not contain the standard sets of organs and do not develop in the same way as the true segments. The frontmost section, called the prostomium contains the brain and sense organs, while the rearmost, called the pygidium or periproct contains the anus, generally on the underside. The first section behind the prostomium, called the peristomium, is regarded by some zoologists as not a true segment, but in some polychaetes the peristomium has chetae and appendages like those of other segments.

The segments develop one at a time from a growth zone just ahead of the pygidium, so that an annelid's youngest segment is just in front of the growth zone while the peristomium is the oldest. This pattern is called teloblastic growth. Some groups of annelids, including all leeches, have fixed maximum numbers of segments, while others add segments throughout their lives.

The phylum's name is derived from the Latin word *annelus*, meaning "little ring".

Body wall, Chetae and Parapodia

Annelids' cuticles are made of collagen fibers, usually in layers that spiral in alternating directions so that the fibers cross each other. These are secreted by the one-cell deep epidermis (outermost skin layer). A few marine annelids that live in tubes lack cuticles, but their tubes have a similar structure, and mucus-secreting glands in the epidermis protect their skins. Under the epidermis is the dermis, which is made of connective tissue, in other words a combination of cells and non-cellular materials such as collagen. Below this are two layers of muscles, which develop from the lining of the coelom (body cavity): circular muscles make a segment longer and slimmer when they contract, while under them are longitudinal muscles, usually four distinct strips, whose contractions make the segment shorter and fatter. Some annelids also have oblique internal muscles that connect the underside of the body to each side.

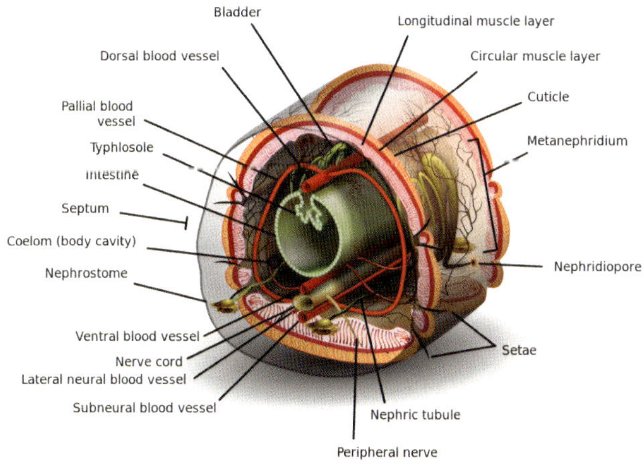

The setae ("hairs") of annelids project out from the epidermis to provide traction and other capabilities. The simplest are unjointed and form paired bundles near the top and bottom of each side of each segment. The parapodia ("limbs") of annelids that have them often bear more complex chetae at their tips – for example jointed, comb-like or hooked. Chetae are made of moderately flexible β-chitin and are formed by follicles, each of which has a chetoblast ("hair-forming") cell at the bottom and muscles that can extend or retract the cheta. The chetoblasts produce chetae by forming microvilli, fine hair-like extensions that increase the area available for secreting the cheta. When the cheta is complete, the microvilli withdraw into the chetoblast, leaving parallel tunnels that run almost the full length of the cheta. Hence annelids' chetae are structurally different from the setae ("bristles") of arthropods, which are made of the more rigid α-chitin, have a single internal cavity, and are mounted on flexible joints in shallow pits in the cuticle.

Nearly all polychaetes have parapodia that function as limbs, while other major annelid groups lack them. Parapodia are unjointed paired extensions of the body wall, and their muscles are derived from the circular muscles of the body. They are often supported internally by one or more large, thick chetae. The parapodia of burrowing and tube-dwelling polychaetes are often just ridges whose tips bear hooked chetae. In active crawlers and swimmers the parapodia are often divided into large upper and lower paddles on a very short trunk, and the paddles are generally fringed with chetae and sometimes with cirri (fused bundles of cilia) and gills.

Nervous System and Senses

The brain generally forms a ring round the pharynx (throat), consisting of a pair of ganglia (local control centers) above and in front of the pharynx, linked by nerve cords either side of the pharynx to another pair of ganglia just below and behind it. The brains of polychaetes are generally in the prostomium, while those of clitellates are in the peristomium or sometimes the first segment behind the peristomium. In some very mobile and active polychaetes the brain is enlarged and more complex, with visible hindbrain, midbrain and forebrain sections. The rest of the central nervous system, the ventral nerve cord, is generally "ladder-like", consisting of a pair of nerve cords that run through the bottom part of the body and have in each segment paired ganglia linked by a transverse connection. From each segmental ganglion a branching system of local nerves runs into the body wall and then encircles the body. However, in most polychaetes the two main nerve cords are fused, and in the tube-dwelling genus *Owenia* the single nerve chord has no ganglia and is located in the epidermis.

As in arthropods, each muscle fiber (cell) is controlled by more than one neuron, and the speed and power of the fiber's contractions depends on the combined effects of all its neurons. Vertebrates have a different system, in which one neuron controls a group of muscle fibers. Most annelids' longitudinal nerve trunks include giant axons (the output signal lines of nerve cells). Their large diameter decreases their resistance, which allows them to transmit signals exceptionally fast. This enables these worms to withdraw rapidly from danger by shortening their bodies. Experiments have shown that cutting the giant axons prevents this escape response but does not affect normal movement.

The sensors are primarily single cells that detect light, chemicals, pressure waves and contact, and are present on the head, appendages (if any) and other parts of the body. Nuchal ("on the neck") organs are paired, ciliated structures found only in polychaetes, and are thought to be chemosensors. Some polychaetes also have various combinations of ocelli ("little eyes") that detect the direction from which light is coming and camera eyes or compound eyes that can probably form images. The compound eyes probably evolved independently of arthropods' eyes. Some tube-worms use ocelli widely spread over their bodies to detect the shadows of fish, so that they can quickly withdraw into their tubes. Some burrowing and tube-dwelling polychaetes have statocysts (tilt and balance sensors) that tell them which way is down. A few polychaete genera have on the undersides of their heads palps that are used both in feeding and as "feelers", and some of these also have antennae that are structurally similar but probably are used mainly as "feelers".

Coelom, Locomotion and Circulatory System

Most annelids have a pair of coelomata (body cavities) in each segment, separated from other segments by septa and from each other by vertical mesenteries. Each septum forms a sandwich with connective tissue in the middle and mesothelium (membrane that serves as a lining) from the preceding and following segments on either side. Each mesentery is similar except that the mesothelium is the lining of each of the pair of coelomata, and the blood vessels and, in polychaetes, the main nerve cords are embedded in it. The mesothelium is made of modified epitheliomuscular cells; in other words, their bodies form part of the epithelium but their bases extend to form muscle fibers in the body wall. The mesothelium may also form radial and circular muscles on the septa, and circular muscles around the blood vessels and gut. Parts of the mesothelium, especially on the outside of the gut, may also form chloragogen cells that perform similar functions to the livers of vertebrates: producing and storing glycogen and fat; producing the oxygen-carrier hemoglobin; breaking down proteins; and turning nitrogenous waste products into ammonia and urea to be excreted.

Many annelids move by peristalsis (waves of contraction and expansion that sweep along the body), or flex the body while using parapodia to crawl or swim. In these animals the septa enable the circular and longitudinal muscles to change the shape of individual segments, by making each segment a separate fluid-filled "balloon". However, the septa are often incomplete in annelids that are semi-sessile or that do not move by peristalsis or by movements of parapodia – for example some move by whipping movements of the body, some small marine species move by means of cilia (fine muscle-powered hairs) and some burrowers turn their pharynges (throats) inside out to penetrate the sea-floor and drag themselves into it.

The fluid in the coelomata contains coelomocyte cells that defend the animals against parasites and infections. In some species coelomocytes may also contain a respiratory pigment – red hemoglobin in some species, green chlorocruorin in others (dissolved in the plasma) – and provide oxygen transport within their segments. Respiratory pigment is also dissolved in the blood plasma. Species with well-developed septa generally also have blood vessels running all long their bodies above and below the gut, the upper one carrying blood forwards while the lower one carries it backwards. Networks of capillaries in the body wall and around the gut transfer blood between the main blood vessels and to parts of the segment that need oxygen and nutrients. Both of the major vessels, especially the upper one, can pump blood by contracting. In some annelids the forward end of the upper blood vessel is enlarged with muscles to form a heart, while in the forward ends of many earthworms some of the vessels that connect the upper and lower main vessels function as hearts. Species with poorly developed or no septa generally have no blood vessels and rely on the circulation within the coelom for delivering nutrients and oxygen.

However, leeches and their closest relatives have a body structure that is very uniform within the group but significantly different from that of other annelids, including other members of the Clitellata. In leeches there are no septa, the connective tissue layer of the body wall is so thick that it occupies much of the body, and the two coelomata are widely separated and run the length of the body. They function as the main blood vessels, although they are side-by-side rather than upper and lower. However, they are lined with mesothelium, like the coelomata and unlike the blood vessels of other annelids. Leeches generally use suckers at their front and rear ends to move like inchworms. The anus is on the upper surface of the pygidium.

Respiration

In some annelids, including earthworms, all respiration is via the skin. However, many polychaetes and some clitellates (the group to which earthworms belong) have gills associated with most segments, often as extensions of the parapodia in polychaetes. The gills of tube-dwellers and burrowers usually cluster around whichever end has the stronger water flow.

Feeding and Excretion

Lamellibrachian tube worms have no gut and gain nutrients from chemoautotrophic bacteria living inside them.

Feeding structures in the mouth region vary widely, and have little correlation with the animals' diets. Many polychaetes have a muscular pharynx that can be everted (turned inside out to extend it). In these animals the foremost few segments often lack septa so that, when the muscles in these segments contract, the sharp increase in fluid pressure from all these segments everts the pharynx very quickly. Two families, the Eunicidae and Phyllodocidae, have evolved jaws, which can be used for seizing prey, biting off pieces of vegetation, or grasping dead and decaying matter. On the other hand, some predatory polychaetes have neither jaws nor eversible pharynges. Selective deposit feeders generally live in tubes on the sea-floor and use palps to find food particles in the sediment and then wipe them into their mouths. Filter feeders use "crowns" of palps covered in cilia that wash food particles towards their mouths. Non-selective deposit feeders ingest soil or marine sediments via mouths that are generally unspecialized. Some clitellates have sticky pads in the roofs of their mouths, and some of these can evert the pads to capture prey. Leeches often have an eversible proboscis, or a muscular pharynx with two or three teeth.

The gut is generally an almost straight tube supported by the mesenteries (vertical partitions within segments), and ends with the anus on the underside of the pygidium. However, in members of the tube-dwelling family Siboglinidae the gut is blocked by a swollen lining that houses symbiotic bacteria, which can make up 15% of the worms' total weight. The bacteria convert inorganic matter – such as hydrogen sulfide and carbon dioxide from hydrothermal vents, or methane from seeps – to organic matter that feeds themselves and their hosts, while the worms extend their palps into the gas flows to absorb the gases needed by the bacteria.

Annelids with blood vessels use metanephridia to remove soluble waste products, while those without use protonephridia. Both of these systems use a two-stage filtration process, in which fluid and waste products are first extracted and these are filtered again to re-absorb any re-usable materials while dumping toxic and spent materials as urine. The difference is that protonephridia combine both filtration stages in the same organ, while metanephridia perform only the second filtration and rely on other mechanisms for the first – in annelids special filter cells in the walls of the blood vessels let fluids and other small molecules pass into the coelomic fluid, where it circulates to the metanephridia. In annelids the points at which fluid enters the protonephridia or metanephridia are on the forward side of a septum while the second-stage filter and the nephridiopore (exit opening in the body wall) are in the following segment. As a result, the hindmost segment (before the growth zone and pygidium) has no structure that extracts its wastes, as there is no following segment to filter and discharge them, while the first segment contains an extraction structure that passes wastes to the second, but does not contain the structures that re-filter and discharge urine.

Reproduction and Life Cycle

- Asexual Reproduction

Polychaetes can reproduce asexually, by dividing into two or more pieces or by budding off a new individual while the parent remains a complete organism. Some oligochaetes, such as *Aulophorus furcatus*, seem to reproduce entirely asexually, while others reproduce asexually in summer and sexually in autumn. Asexual reproduction in oligochaetes is always by dividing into two or more pieces, rather than by budding. However, leeches have never been seen reproducing asexually.

This sabellid tubeworm is budding

Most polychaetes and oligochaetes also use similar mechanisms to regenerate after suffering damage. Two polychaete genera, *Chaetopterus* and *Dodecaceria*, can regenerate from a single segment, and others can regenerate even if their heads are removed. Annelids are the most complex animals that can regenerate after such severe damage. On the other hand, leeches cannot regenerate.

- Sexual Reproduction

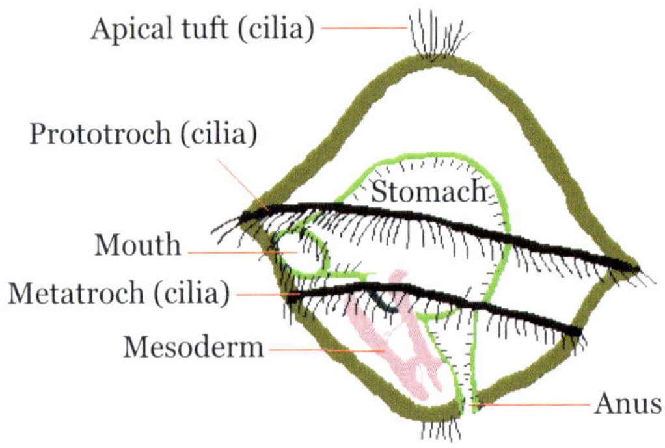

Trochophore larva

It is thought that annelids were originally animals with two separate sexes, which released ova and sperm into the water via their nephridia. The fertilized eggs develop into trochophore larvae, which live as plankton. Later they sink to the sea-floor and metamorphose into miniature adults: the part of the trochophore between the apical tuft and the prototroch becomes the prostomium (head); a small area round the trochophore's anus becomes the pygidium (tail-piece); a narrow band immediately in front of that becomes the growth zone that produces new segments; and the rest of the trochophore becomes the peristomium (the segment that contains the mouth).

However, the lifecycles of most living polychaetes, which are almost all marine animals, are unknown, and only about 25% of the 300+ species whose lifecycles are known follow this pattern. About 14% use a similar external fertilization but produce yolk-rich eggs, which reduce the time the larva needs to spend among the plankton, or eggs from which miniature adults emerge rather than larvae. The rest care for the fertilized eggs until they hatch – some by producing jelly-covered masses of eggs which they tend, some by attaching the eggs to their bodies and a few species by keeping the eggs within their bodies until they hatch. These species use a variety of methods for sperm transfer; for example, in some the females collect sperm released into the water, while in others the males have a penis that inject sperm into the female. There is no guarantee that this is a representative sample of polychaetes' reproductive patterns, and it simply reflects scientists' current knowledge.

Some polychaetes breed only once in their lives, while others breed almost continuously or through several breeding seasons. While most polychaetes remain of one sex all their lives, a significant percentage of species are full hermaphrodites or change sex during their lives. Most polychaetes whose reproduction has been studied lack permanent gonads, and it is uncertain how they produce ova and sperm. In a few species the rear of the body splits off and becomes a separate individual that lives just long enough to swim to a suitable environment, usually near the surface, and spawn.

Most mature clitellates (the group that includes earthworms and leeches) are full hermaphrodites, although in a few leech species younger adults function as males and become female at maturity. All have well-developed gonads, and all copulate. Earthworms store their partners' sperm in spermathecae ("sperm stores") and then the clitellum produces a cocoon that collects ova from the ovaries and then sperm from the spermathecae. Fertilization and development of earthworm eggs takes place in the cocoon. Leeches' eggs are fertilized in the ovaries, and then transferred to the cocoon. In all clitellates the cocoon also either produces yolk when the eggs are fertilized or nutrients while they are developing. All clitellates hatch as miniature adults rather than larvae.

Mollusca

Mollusca is one of the most diverse groups of animals on the planet, with at least 50,000 living species (and more likely around 200,000). It includes such familiar organisms as snails, octopuses, squid, clams, scallops, oysters, and chitons. Mollusca also includes some lesser known groups like the monoplacophorans, a group once thought to be extinct for millions of years until one was found in 1952 in the deep ocean off the coast of Costa Rica.

Molluscs are a clade of organisms that all have soft bodies which typically have a "head" and a "foot" region. Often their bodies are covered by a hard exoskeleton, as in the shells of snails and clams or the plates of chitons.

A part of almost every ecosystem in the world, molluscs are extremely important members of many ecological communities. They range in distribution from terrestrial mountain tops to the hot vents and cold seeps of the deep sea, and range in size from 20-meter-long giant squid to microscopic aplacophorans, a millimeter or less in length, that live between sand grains.

These creatures have been important to humans throughout history as a source of food, jewelry, tools, and even pets. For example, on the Pacific coast of California, Native Americans consumed large quantities of abalone and especially owl limpets. However, the impact of Native Americans on these molluscan communities pales by comparison to the overharvesting of some molluscan taxa by the United States in the 1960s and 1970s. Species whose members once numbered in the millions, now teeter on the verge of extinction. For example, fewer than 100 white abalone remain after several million individuals were captured and sold as meat in the 1970s. Besides having yummy soft parts, molluscs often have desirable hard parts. The shells of some molluscs are considered quite beautiful and valuable. Molluscs can also be nuisances, such as the common garden snail; and molluscs make up a major component of fouling communities both on docks and on the hulls of ships.

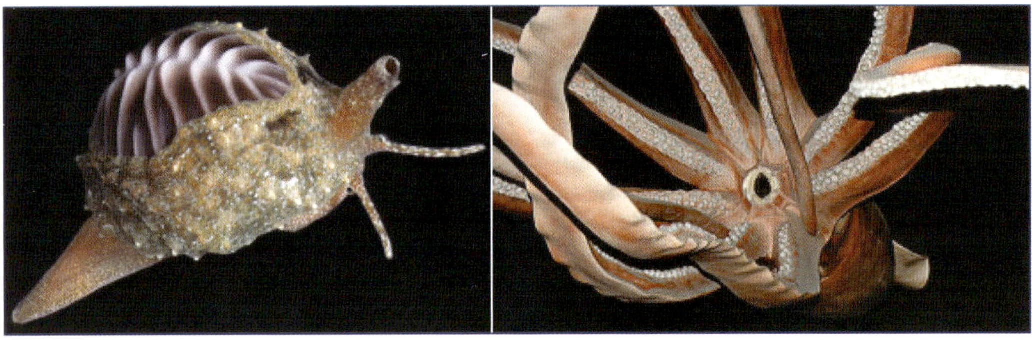

On the left is a marine snail, the California Trivia (*Trivia californiana*). Here the mantle covers much of the shell. Note how a portion of the mantle is rolled into a tube shape to form the siphon just above the head. At the right is a restoration of one of the largest of all molluscs, the Giant Squid (*Architeuthis*).

They also have a very long and rich fossil record going back more than 550 million years, making them one of the most common types of organism used by paleontologists to study the history of life.

Systematics

Molluscan systematics are still in flux. As you can see from the cladogram below, there is still no agreement on some of the major relationships. The polytomies shown indicate that the question of which molluscs are the most closely related is still a matter of debate.

However, new types of data and much larger and more sophisticated analyses continue to be performed. The resolved relationships shown (such as cephalopods, scaphopods, and gastropods) are recent discoveries.

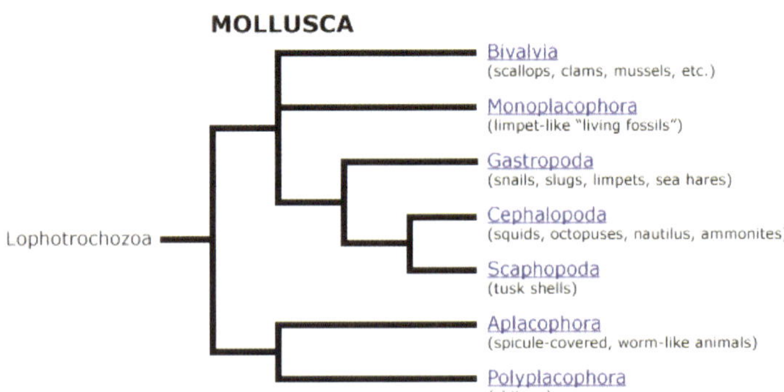

Morphology

Despite their amazing diversity, all molluscs share some unique characteristics that define their body plan. The body has a head, a foot and a visceral mass. This is all covered with a mantle (also known as a pallium) that typically secretes the shell. In some groups, like slugs and octopuses, the mantle is secondarily lost, while in others, it is used for other activities, such as respiration.

The freshwater Sinistral Pond Snail (*Physella* sp.) scrapes algae from the glass with its radula, the two "toothy" arcs you can see lining the mouth.

The buccal cavity, at the anterior of the mollusc, contains a radula (lost in bivalves) — a ribbon of teeth supported by an odontophore, a muscular structure. The radula is generally used for feeding. The ventral foot is used in locomotion. This foot propels the mollusc by utilizing muscular waves and/or cilia in combination with mucus.

Typically, at least in the more primitive members of each group, there are one or more pairs of gills (called ctenidia) which lie in a posterior cavity (the pallial cavity) or in a posterolateral groove surrounding the foot. The pallial cavity typically contains a pair of sensory osphradia (for smelling) and is the space into which the kidneys, gonads, and anus open.

Molluscs are coelomate, although the coelom is reduced and represented by the kidneys, gonads, and pericardium, the main body cavity which surrounds the heart.

Life History and Ecology

Molluscs occur in almost every habitat found on Earth, where they are often the most conspicuous organisms. While most are found in the marine environment, extending from the intertidal to the deepest oceans, several major gastropod clades live predominantly in freshwater or terrestrial habitats. Remarkably, one study found around 3000 species within a single locality at a coral reef in New Caledonia. In terrestrial communities, gastropods can achieve reasonably high diversity and abundance: as many as 60-70 species may coexist in a single habitat and abundance in leaf litter can exceed more than 500 individuals in four liters of litter.

Many marine molluscs emerge from their eggs as planktonic trochophore larvae, however, Sinistral Pond Snails (*Physella* sp.) emerge from their eggs as young snails. The whitish, jellybean-shaped organisms are ostracodes (crustaceans).

Marine molluscs occur on a large variety of substrates including rocky shores, coral reefs, mud flats, and sandy beaches. Gastropods and chitons are characteristic of these hard substrates, and bivalves are commonly associated with softer substrates where they burrow into the sediment. However, there are many exceptions: the largest living bivalve, *Tridacna gigas*, lives on coral reefs, and many bivalves (e.g., mussels and oysters) attach themselves to hard substrates. Some microscopic gastropods even live interstitially between sand grains.

Large concentrations of gastropods and bivalves are found at hydrothermal vents in the deep sea. Living in these or other dysoxic habitats appears to be a plesiomorphic condition for the Mollusca and several outgroups. For example, the fauna of Palaeozoic hydrothermal vent communities includes the molluscan groups Bivalvia, Monoplacophora and Gastropoda as well as the outgroups Brachiopoda and Annelida.

The adoption of different feeding habits appears to have had a profound influence on molluscan evolution. The change from grazing to other forms of food acquisition is one of the major features in the radiation of the group. Based on our current understanding of relationships, the earliest molluscs grazed on encrusting animals and detritus. Such feeding may have been selective or indiscriminate and will have encompassed algal, diatom, or cyanobacterial films and mats, or encrusting colonial animals. Truly herbivorous grazers are relatively rare and are limited to some polyplacophorans and a few gastropod groups. Most chaetodermomorph aplacophorans, monoplacophorans and scaphopods feed on protists and/or bacteria while neomeniomorph aplacophorans graze on cnidarians. Cephalopods are mainly active predators as are some gastropods, while a few chitons and septibranch bivalves capture microcrustaceans. Most bivalves are either suspension or deposit feeders that indiscriminately take in particles, but then elaborately sort them based on size and weight, typically assimilating bacteria, protists, and diatoms.

References

- Liem, K.F.; Walker, W.F. (2001). Functional anatomy of the vertebrates: an evolutionary perspective. Harcourt College Publishers. p. 277. ISBN 978-0-03-022369-3
- Harlow, H. F.; Suomi, S. J. (1971). "Social Recovery by Isolation-Reared Monkeys". Proceedings of the National Academy of Sciences of the United States of America. 68 (7): 1534–1538. Bibcode:1971PNAS...68.1534H. doi:10.1073/pnas.68.7.1534. PMC 389234
- Reich, Ronni (15 October 2010). "NJIT professor finds nothing cuckoo in serenading our feathered friends". Star Ledger. Retrieved 19 June 2011
- Teng, L.; Labosky, P. A. (2006). "Neural crest stem cells" In: Jean-Pierre Saint-Jeannet, Neural Crest Induction and Differentiation, pp. 206-212, Springer Science & Business Media. ISBN 9780387469546
- "Tuna groups tackle overfishing". BBC News. 26 January 2007. Archived from the original on 21 January 2009. Retrieved 18 January 2006
- Nicholls, J.G. and Baylor, D.A., (1968). Specific modalities and receptive fields of sensory neurons in CNS of the leech. Journal of Neurophysiology, 31: 740–756
- Andersen, N.M.; Weir, T.A. (2004). Australian water bugs: their biology and identification (Hemiptera-Heteroptera, Gerromorpha & Nepomorpha). Apollo Books. p. 38. ISBN 978-87-88757-78-1
- Sahney, S., Benton, M.J. and Ferry, P.A. (2010). "Links between global taxonomic diversity, ecological diversity and the expansion of vertebrates on land" (PDF). Biology Letters. 6 (4): 544–547. doi:10.1098/rsbl.2009.1024. PMC 2936204. PMID 20106856
- Byrne, J.H., Castellucci, V.F. and Kandel, E.R., (1978). Contribution of individual mechanoreceptor sensory neurons to defensive gill-withdrawal reflex in Aplysia. Journal of Neurophysiology, 41: 418–431
- Klopfer, P. H. (1981). "Origins of Parental Care". In Gubernick, D. J. Parental Care in Mammals. New York: Plenum Press. ISBN 978-1-4613-3150-6. OCLC 913709574
- Antcliffe, J.B.; Brasier, Martin D. (2007). "Charnia At 50: Developmental Models For Ediacaran Fronds". Palaeontology. 51(1): 11–26. doi:10.1111/j.1475-4983.2007.00738.x

3

Animal Classification by Feeding Behavior

Animals can be categorized according to their feeding habits into the ecological groups of herbivores, carnivores, omnivores, fungivores and bacterivores. This chapter closely examines these key animal groups in great detail to provide an extensive understanding of their feeding behavior.

Herbivore

An herbivore is an organism that mostly feeds on plants. Herbivores range in size from tiny insects such as aphids to large, lumbering elephants.

Herbivores are a major part of the food web, a description of which organisms eat other organisms in the wild. Organisms in the food web are grouped into trophic, or nutritional, levels. There are three trophic levels. Autotrophs, organisms that produce their own food, are the first trophic level. These include plants and algae. Herbivores, which eat autotrophs, are the second trophic level. Carnivores, organisms that consume animals, and omnivores, organisms that consume both plants and animals, are the third trophic level.

Autotrophs are called producers, because they produce their own food. Herbivores, carnivores, and omnivores are consumers. Herbivores are primary consumers. Carnivores and omnivores are secondary consumers.

Herbivores often have physical features that help them eat tough, fiberous plant matter. Unlike herbivores and other consumers, autotrophs have tough cell walls throughout their physical structure. Cell walls can make plant material difficult to digest.

Many herbivorous mammals have wide molars. These big teeth help them grind up leaves and grasses. Carnivorous mammals, on the other hand, usually have long, sharp teeth that help them grab prey and rip it apart.

A group of herbivores called ruminants have specialized stomachs. For the digestion of plant matter, ruminant stomachs have more than one chamber. When a ruminant chews up and swallows grass, leaves, and other material, it goes into the first chamber of its stomach, where it sits and softens. There, specialized bacteria break down the food. When the material is soft enough, the animal regurgitates the food and chews it again. This helps break down the plant matter. This partially digested food is called cud. The animal then swallows the cud, and it goes into a second chamber of the stomach. Chemicals in the second chamber digest the plant material further, and it goes into the third chamber. Finally, the digested food goes to the fourth chamber, which is similar to a human stomach. Sheep, deer, giraffes, camels, and cattle are all ruminants.

Picky Eaters

Some herbivores eat any plant matter they can find. Elephants, for example, eat bark, leaves, small branches, roots, grasses, and fruit. Black rhinoceroses also eat a variety of fruits, branches, and leaves.

Other herbivores eat only one part of a plant. An animal that specializes in eating fruit is called a frugivore. Oilbirds, which live in northern South America, are frugivores. They eat nothing but the fruit of palms and laurels. The koala, which is native to Australia, eats little besides the leaves of eucalyptus trees. An animal that eats the leaves and shoots of trees is called a folivore. Pandas, which feed almost exclusively on bamboo, are folivores. Termites are insects that feed mostly on wood. Wood-eaters are called xylophages.

Many insects are herbivores. Some, such as grasshoppers, will eat every part of a plant. Others specialize in certain parts of the plant. Aphids drink sap, a sticky fluid that carries nutrients through the plant. Caterpillars eat leaves. The larvae, or young wormlike forms, of root weevils feed on roots. Asian long-horned beetles tunnel deep into the heart of a tree and eat the wood there. Honeybees feed on nectar and pollen from flowers.

Some herbivores consume only dead plant material. These organisms are called detritivores. Detritivores also consume other dead organic material, such as decaying animals, fungi, and algae. Detritivores such as earthworms, bacteria, and fungi are an important part of the food chain. They break down the dead organic material and recycle nutrients back into the ecosystem. Detritivores can survive in many places. Earthworms and mushrooms live in the soil. There are also detritivore bacteria at the bottom of the ocean.

Plants that are parasites can still be considered herbivores. A parasite is an organism that lives on or in another organism and gets its nutrients from it. Parasitic plants get their nutrients from other plants, called host plants. Dodder, native to tropical and temperate climates around the world, is a parasitic vine that wraps around a host plant. Dodder has rootlike parts called haustoria that attach to the host plant, so it can feed on its nutrients. Eventually, the parasitic dodder feeds on all the nutrients of the host plant, and the host plant dies. The dodder vines then move on to another plant.

Plant/Herbivore Relationships

Herbivory is an interaction in which a plant or portions of the plant are consumed by an animal. At the microscopic scale, herbivory includes the bacteria and fungi that cause disease as they feed on plant tissue. Microbes that break down dead plant tissue are also specialized herbivores. Browsers and grazers, from aphids and caterpillars to deer and bison, are more familiar herbivores. Even insects and animals that eat seeds are considered herbivores.

Some herbivores consume entire plants, or enough to kill them. Others only eat a portion of the plant, and so the plant can recover. The plant/herbivore relationship traditionally has been seen as lopsided, with the animal as the beneficiary and the plant as the loser. Current research, however, is revealing that herbivory has some potential benefits to plants. One example is canopy grazing by insects, which allows more light to penetrate into the lower layers of the forest. Gypsy moth grazing on canopy trees in some areas of Virginia's Blue Ridge Mountains, for instance, has resulted in more light penetration and therefore a more diverse and productive ground layer.

Herbivores and their Food Plants

Bison, sheep, and other grazers - Succulent forbs, grasses, grass-like plants

Deer and other ungulate browsers - Leaves and twigs of woody plants such as willows, arborvitaes, yews

Beaver - Tree bark, young shoots, leaves

Rodents - Succulent forbs, grasses, grass-like plants

Rabbits - Succulent forbs, grasses, bark

Voles - Roots, bark

Caterpillars - Leaves; in some cases, of specific species

Monarch butterfly - Milkweeds

Gypsy moth - Oaks and other hardwoods

Aphids - Plant juices; in some cases, of specific species

Many birds - Seeds and fruits

Locusts - All plants; seeds, leaves, and stems

Herbivory is the consumption of plant tissues by animals. This usually has a negative impact on plant growth and reproduction, and so imposes natural selection on plants, thereby favoring the evolution of traits that reduce losses (defenses).

Plants have developed physical defenses that reduce the likelihood of herbivory.

Terrestrial plants generally lose less than 10 percent of annual production to herbivores. Fish can consume 100 percent in some shallow-water marine systems in days or weeks, and single-celled algae are often eaten as fast as they reproduce. A few terrestrial herbivores (gypsy moths or locusts, for example) occasionally experience population explosions during which they may totally defoliate their preferred food plants over wide areas A loss of even 15 percent of annual production can reduce plant growth and fitness.

The fact that some herbivores occasionally can consume most or all vegetation suggests that something prevents this most of the time. Two dominant opposing hypotheses are that "top-down" forces (predators, parasites , disease) limit herbivore populations so all plants are not consumed, or that "bottom-up" forces (plant quality) prevent herbivores from eating some plant tissues. In reality both forces combine to limit herbivores and herbivory. Herbivore damage is greater in simplified managed systems, reflecting the loss of these interactions.

Plants have developed numerous physical (spines, thorns, tough tissues, sticky resins, and hairs) and chemical defenses that may reduce herbivory. Some of these are fixed (constitutive), whereas others are only produced (induced) when the plant is attacked. A tremendous diversity of plant biochemicals are toxic, repellent, or antinutritive for herbivores of all types. Many of these chemicals have been called "secondary metabolites "because roles have not been found for them in primary plant functions like growth and reproduction. Examples include alkaloids (such as caffeine and nicotine), terpenoids (terpene and pinene), glucosinolates (sinigrin), and phenolics (tannin and chlorogenic acid).

Although many of these compounds appear to have specific activity in animal systems, such as interfering with neurotransmission, recent research suggests that they may have other functions in plants. Nonetheless, it is clear that plant chemistry is a major barrier to herbivory. It is also exploited by humans in medicinal plant use and pharmaceutical development.

While all plants produce chemical defenses continuously, all plants so far studied also change or increase production of both physical and chemical defenses when attacked by herbivores. Defenses can be induced throughout a plant, even in unattacked tissues or tissues produced after the attack, producing systemic resistance. New physical and chemical defenses may be synthesized, enzymes may activate preexisting defenses, or tissues may be dropped (abscised) to remove pests. This requires detection, coordination, and response. Plants can discriminate among and respond differentially to wounding (such as wind damage) and herbivores, or even among herbivores.

Chemicals found in insect regurgitant (mouth juices) trigger plant responses; these cues may be produced by the insect or by bacteria living in their guts. Fatty acid signals made by the plant (especially jasmonic acid) in response to attack circulate widely, stimulating defense gene expression and providing systemic resistance. Methyl jasmonate is volatile and escapes wounded plants, triggering defense responses in nearby unwounded plants.

Because producing defenses requires materials (such as carbon and nitrogen) and energy that presumably could be used for growth or reproduction, many believe that defense may be costly for plants. Clear evidence of this is difficult to obtain, but some plants grow or reproduce less when producing maximum defenses, providing indirect support. The types of defenses employed by fast-growing versus slow-growing, or early-versus late-successional plant species differ in consistent ways, but why this is so is not clear. One benefit of induced defenses could be cost-saving, since they are only produced when needed. Induced or constitutive, plant defense chemistry influences litter quality, decomposition and nutrient cycling in ecosystems .

Herbivores, especially insects, have developed behavioral and biochemical mechanisms that reduce the effects of plant defenses. Plant evolution includes ongoing development of new defenses, which is thought to favor the evolution of new herbivore adaptations and promote speciation in both plants and herbivores. This kind of reciprocal evolutionary impact is called coevolution. While all plants gain some protection from defenses, no plant escapes herbivory by at least some adapted herbivores. Plant defenses and consumer adaptation can limit consumption, and thus can determine the length and shape of food webs .

Feeding Strategies

Herbivores consume many different parts of plants. This group of animals can be subdivided into folivores (eaters of leaves), frugivores (eaters of fruit), and granivores (seed eaters).

Caterpillars are probably the most familiar example of folivores. Caterpillars must eat large quantities of leaves in order to store enough fat and protein to undergo metamorphosis. However, some plants have developed a defense against caterpillars. In Finland, caterpillars of the moth *Oporinia autumnata* eat leaves of the birch tree. Caterpillars that feed on leaves from trees that were heavily damaged the previous year grow more slowly than caterpillars that feed on lightly damaged trees. Apparently the birch tree has developed chemicals that make the damaged leaves hard to digest.

Species of milkweed have evolved a highly toxic sap that prevents most caterpillars from consuming their leaves. However, the caterpillar of the monarch butterfly has evolved a tolerance for the toxin in milkweed sap and readily consumes it. Toxins from the sap are incorporated into the tissue of the caterpillars, making the caterpillars unpalatable to most predators. However, certain species of orioles have evolved a tolerance for this toxin and readily consume the insect.

Mammalian grazers generally consume a variety of different plants during a day. Only a few, such as the koala of Australia, eat only one type. Large species of grazers tend to be less selective than smaller species. In East Africa, communities of grazing animals have developed a commensal relationship. Elephants and buffalo first eat the tall, coarse grasses and then move on. They are able to consume large quantities of this low-nutrition food source. Zebras follow along behind the elephants, reducing the plant biomass even more. The zebras are followed by still smaller wildebeest, who select among the lower-growing plants that remain after the zebras have fed. Finally the smallest grazers such as Thompson's gazelles are able to reach the young, protein-rich sprouts of grass missed by the wildebeest.

Grazing

Grazers crop grasses and other ground plants on land or scrape algae and other organisms from surfaces in the water. They include animals as diverse as snails, grasshoppers, geese, rodents, kangaroos, and hoofed mammals. Grass and algae are palatable foods that offer little or no resistance to being eaten, but are adapted to survive grazing and quickly replace the lost biomass. A disadvantage of such food, however, is that it is nutrient poor. Grazers therefore must consume a large quantity of it and spend a larger percentage of their time eating than predators do. While eating, they are vulnerable to attack. To eat without being eaten requires alertness and quick escape responses. Grazing mammals tend to form herds: There is safety in numbers, and the abundance of grass supports the high population density of grazing herds.

Browsing

Terrestrial browsers nip foliage from trees and shrubs. They include caterpillars, tortoises, grouse, giraffes, goats, antelopes, deer, pandas, koalas, and monkeys. In aquatic habitats, browsers feed on algae, aquatic plants, and corals, and include sea slugs, sea urchins, parrot fish, ducks, and manatees. Browsers depend on food that is less abundant and widespread than grass, so they tend to form smaller groups or to be solitary and secretive.

Eating Nectar, Fruits, Pollen, and Seeds

Plants provide an abundance of food other than foliage, some of it for the purpose of rewarding animals. Sweet nectar rewards bees, flies, moths, butterflies, and bats that spread pollen from one flower to another, and sugary fruits entice birds, monkeys, fruit bats, bears, elephants, and humans to eat them and spread the indigestible seeds throughout the countryside. Pollen and seeds, being a plant's reproductive capital, are not meant to be eaten, but many bees, flies, and beetles nevertheless consume pollen, while birds, squirrels, and harvester ants take their toll on the seed crop.

Burrowing

Some animals burrow into their food, eating a tunnel as they go. These include many herbivores such as bark beetles, fly and moth larvae called leaf miners, and wood-boring termites. In the sea, unusual clams and crustaceans called shipworms and gribbles, respectively, burrow through wooden piers and ships, causing enormous destruction. Earthworms and many marine worms burrow in soil and sediment, eating indiscriminately as they go, digesting the organic matter and defecating the indigestible sand and other particles. Burrowing animals not only have the benefit of being surrounded by food, but also are less exposed to predators.

Filter-feeding

Filter-feeding is a common strategy in aquatic habitats, especially the ocean. It uses anatomical devices that act as strainers to remove small food items from the water. Sessile filter-feeders, such as barnacles, oysters, fanworms, brachiopods, and tunicates sit in one place, pumping sea water and straining plankton from it. Other filter-feeders are mobile. Herring swim with their mouths open, letting water flow through the gill rakers, which strain small particles of food from it. Flamingoes take in mouthfuls of water and mud, then force the water through the fringed edges of their bills, which serve as strainers that retain food such as brine shrimp, aquatic insects, and plankton in the mouth. Small and even microscopic food in the water may not seem very abundant, yet the largest animals on Earth—the basking sharks, whale sharks, manta rays, and baleen whales, including the largest species alive today, the great blue whale—nourish themselves entirely in this way. Filter-feeding is more common in the ocean than in fresh water, because plankton is less concentrated in fresh water.

Suspension and Deposit Feeding

Another form of small particulate food in aquatic habitats is the steady "rain" of organic matter that settles to the bottom: living and dead plankton and bits of dead animal, plant, and algal tissue. Suspension feeders pick this material from the water as it falls and deposit feeders consume it after it settles on the bottom. Many sea anemones, corals, marine worms, and crinoids, for example, spread out an array of tentacles and capture whatever settles on them. Other worms, some bivalves, brittle stars, and sea cucumbers spread sticky palps, arms, or tentacles over the substrate , picking up the organic matter that has settled there. The feeding arms or tentacles of many of these animals have ciliated , sticky grooves. Food becomes caught in mucus, and cilia steadily propel the mucus strand toward the mouth. Sea cucumbers, however, reach out and pick up sediment on their sticky tentacles, then draw the tentacles into their mouths and remove the food, like licking jam off one's fingers.

Predation

Predators are animals that depend on killing other animals outright. Since the other animals have evolved defenses against predation—hard shells, toxins, the ability to fight back, or simply running or flying away—predators have evolved a wide range of strategies for capturing their prey. Some hunt in packs (wolves), some collaborate to ambush prey (lions), some are stalkers (solitary cats), some use lures to attract unsuspecting prey (snapping turtles and angler fish), some employ camouflage so their prey does not notice them until it is too late (praying mantids), and some use snares (spiders, jellyfish).

Symbiosis

Symbionts are animals that live in a close physical relationship with another animal, the host, from which they benefit. Unlike predators, symbionts do not benefit from the death of their hosts; ideally, they steal food or consume host tissue at a rate that the host can tolerate, allowing the host to survive. Symbiosis includes mutually beneficial relationships (mutualism); relationships in which one partner benefits, typically by stealing food from the host or eating its tissues, but the host is neither benefited nor harmed (commensalism); and relationships in which the host is harmed, usually because the symbiont consumes nutrients or tissue faster than the host can replace it (parasitism). The host is often both food and shelter for its symbiont.

Scavenging

Finally, and fortunately for the planet's "hygiene," many animals belong to a community of scavengers that feed on organic refuse such as manure (dung beetles, flies), leaf litter (snails, millipedes, earthworms), and dead animals (blowflies, vultures, hyenas, storks). The family name of the vultures, Cathartidae, is from the Greek katharos, meaning "to cleanse." Disgusting as some people may find their habits, we would be infinitely more disgusted with an environment from which such scavengers were lacking.

Folivore

The leaf eaters of the world, folivores including such far-out creatures as the giraffelike okapi, Australia's eucalyptus-eating koala , the moss-green owl parrot of New Zealand, the indri lemur of Madagascar, and the slow-moving sloth all must be highly specialized to survive. To eat leaves, which are abundant yet all-but-void of nutrition, most folivores have specialized stomachs, many are slow moving and sleep a lot, and all excel at making a little energy go a long way. It is these energy-saving adaptations, however, which also lock folivores into their own strange world and make them vulnerable to change. For folivores, the good news is that leaves are almost everywhere. There are billions of them, just sitting there, rustling in the breeze. As well as being easy to find, they don't run away nor do they hide. The bad news is that some of the leaves fight back. The war between folivores and trees has become a complicated arms race, with trees trying to defend their leaves and the eaters attempting to overcome the defenses. One line of leaf defense is physical; such as, an array of spines, hairs, wax, and a cellulose-like lignin that make certain leaves difficult to bite or chew. Other plants are so rich in silica that they can wear a chewer's teeth to stubs in no time. Many leaves use chemicals for the second line of defense. These chemicals are aimed mainly at insects that consume most of the world's foliage. Inside the leaves lurks a natural pharmacopeia of toxins, insecticides, and antifungal compounds. Plants produce substances like nicotine, pyrethrin, strychnine, and phenols that provide them with such effective protection that scientists see some of these compounds as alternatives to synthetic insecticides. Folivoric mammals have two major strategies for dealing with plant toxins. Some, like koalas and sloths, have very narrow diets and tend to specialize in detoxifying a few kinds of leaves. Others, such as howler monkeys, often deal with the chemical poisons in their food by eating just a few samples from manydifferent trees. The simplest way to extract energy form leaves is to chew, squash, or grind the leaf to release the digestible cell contents. The process involves considerable chewing and there is little energy reward, but, apart from grinding teeth, no special equipment is needed. To summon any more energy from a diet of leaves requires a special enzyme that can digest the cell walls themselves. Mammals and birds cannot produce this useful digestive juice, but bacteria can. Serious leaf-eaters like sloths, koalas and just about all folivores (except pandas), keep leaf-digesting bacteria in ther guts. Most folivores also have specialized digestive tracts. Sloths and tree kangaroos have large complex stomachs that, like a cow's belly, function as fermentation chambers. Bacteria break down the leaves, but food can take a week or more to pass through the digestive tract. Other folivores such as howler monkeys and indri lemurs have a different strategy: simple stomachs which rely on absorbing nutrients through an extra-long intestine. Sloths, colobine monkeys and some types of antelope have gut bacteria that are finely tuned to digest only selective kinds of leaves. The demands of living on low-energy and often poisonous food means that most folivores live close to the limit of their energy supply.

They must conserve energy wherever possible, and this often translates into being very slow. It is no coincidence that the sloth, the world's slowest mammal, is a folivore. Besides being slow, folivores also spend much of their time resting. Koalas spend 18 to 22 hours a day sleeping, and even when they are awake, they move slowly and look tired. Their lethargy reflects an extremely slow metabolism, which allows them to survive on an energy-poor diet. Sloths, iguanas and possibley other folifores also use solar heat to compensate for low energy levels. For folivorous birds, life is even more difficult. Flying requires an enormous amount of energy, so most birds that feed on leaves have either turned to gliding or abandoned flight completely. Like the flightless owl parrot of New Zealand, folivorous birds tend to be found on islands where there are few natural predators. Because rearing the young is one of life's most energy expensive activities, some leaf-eaters may give birth only once every three or four years. Typically they produce just one small offspring, which grows very slowly and depends on its mother for a long time. The giant panda can ill afford to waste limited energy on its young either. They do not conceive until they are five or six years old and then they only have one or two tiny young. Like the panda, many other folivores find themselves backed into a corner by their highly specialized ways.

Graminivore

Red kangaroo eating grass

In zoology, a graminivore is an herbivorous animal that feeds primarily on grass(specifically "true" grasses, plants of the family Poaceae). The word is derived from Latin *graminis*, meaning "grass", and *vorare*, meaning "to eat." Graminivory is a form of grazing. These herbivorous animals have digestive systems that are adapted to digest large amounts of cellulose, which is abundant in fibrous plant matter and more difficult to break down for many other animals. As such, they have specialized enzymes to aid in digestion and in some cases symbiotic bacteria that live in their digestive track and "assist" with the digestive process through fermentation as the matter travels through the intestines.

Horses, cattle, capybara, hippopotamuses, geese, and giant pandas are examples of vertebrate graminivores. Some carnivorous vertebrates, such as dogs and cats, are known to eat grass occasionally. Grass consumption in dogs can be a way to rid their intestinal tract of parasites that may be threatening to the carnivore's health. Various invertebrates also have graminivorous diets. Many grasshoppers, such as individuals from the Acrididae family, have diets consisting primarily of plants from the Poaceae family.

Horses also eat grass

Graminivores generally exhibit a preference on which species of grass they choose to consume. For example, according to a study done on North American bison (*Bison bison L.)* feeding on shortgrass plains in north-eastern Colorado, the cattle consumed a total of thirty-six different species of plant. Of that thirty-six, five grass species were favoured and consumed the most pervasively. The average consumption of these five species comprised about 80% of their diet. A few of these species include *Aristida longiseta Steud.*, *Muhlenbergia sp.*, and *Bouteloua gracilis (H.B.K.) Lag.*

A capybara grazing at Hattiesburg Zoo

Florivore

In zoology, a florivore is an animal whose diet consists mainly of the products of flowers. Florivores are types of herbivores (often referred to as floral herbivores), yet within the feeding behaviour of florivory, there are a range of other more specific feeding behaviours, including, but not limited to:

- Granivory: the consumption of grain and seeds
- Nectarivory: the consumption of flower nectar
- Palynivory: the consumption of flower pollen
- Frugivory: the consumption of fruit

Diet

A florivore's diet consists of bulky foodstuffs, including the items mentioned above, yet also bark, roots, and similar items. Many florivores are also omnivores, meaning that their diets can also be supplemented by various small insects, for instance.

Examples

The majority of the Psittacine family are florivores, which includes most Parrots, Parakeets, Macaws, and Cockatoos. Other notable florivores are Hummingbirds, Sparrows, and Toucans.

The Parrot: many are florivores, yet some are omnivores.

Granivory

Seed predation, often referred to as granivory, is a type of plant-animal interaction in which granivores (seed predators) feed on the seeds of plants as a main or exclusive food source, in many cases leaving the seeds damaged and not viable. Granivores are found across many families of vertebrates (especially mammals and birds) as well as invertebrates (mainly insects); thus, seed predation occurs in virtually all terrestrial ecosystems. Seed predation is commonly divided into two distinctive temporal categories, pre-dispersal and post-dispersal predation, which may involve different strategies and requirements and have different implications at the individual and population level. Nevertheless, regardless of timing, seed predation has great implications for plant population dynamics. To counterbalance effects of predation, plants have evolved defenses such as seed morphology (size, shape, toughness) and chemical defenses (secondary compounds such as tannins and alkaloids) to defend against their seed predators. However, as plants have adapted defenses to seed predation (e.g., chemical compounds), so have seed predators adapted to plant defenses (e.g., ability to detoxify chemical compounds). Thus, many interesting examples of coevolution arise from this dynamic relationship.

Seeds and their Defenses

Plant seeds are important sources of nutrition for animals across most ecosystems. Seeds contain food storage organs (e.g., endosperm) that provide nutrients to the developing plant embryo (cotyledon). This makes seeds an attractive food source for animals because they are a highly concentrated and localized nutrient source in relation to other plant parts.

Seeds of many plants have evolved a variety of defenses to deter predation from seed predators. Plants may allocate resources toward physical or chemical defenses. Seeds are often contained inside protective structures or fruit pulp that encapsulate seeds until they are ripe. Other physical defenses include spines, hairs, fibrous seed coats and hard endosperm. Seeds, especially in arid areas, may have a mucilaginous seed coat that can glue soil to seed hiding it from granivores.

Some seeds have evolved strong anti-herbivore chemical compounds. In contrast to physical defenses, chemical defenses in seeds to deter consumption by seed predators by using chemicals that are toxic to granivores or inhibit the digestibility of the seed. These chemicals include non-protein amino acids, cyanogenic-glycosides, protease and amylase inhibitors, phytohaemaglutinins. Seeds face trade-offs between allocation toward defenses and the size and number of seeds produced. Additionally plants have trade-offs in seed defense allocation between deterring seed predators and attracting seed dispersers.

Plants may reduce the severity of seed predation by making seeds spatially or temporally scarce to granivores. Seed dispersal away from the parent plant is hypothesized to reduce the severity of seed predation. Seed masting is an example of how plant populations are able to temporally regulate the severity of seed predation. Masting refers to a concerted abundance of seed production followed by a period of paucity. This strategy has the potential to regulate the size of the population of seed predators.

Pre- and Post-dispersal

Seed predation often occurs at contrasting plant ontogenic stages. Predispersal seed predation takes place when seeds are removed from the parent plant before dispersal, and it has been most often reported in invertebrates, birds, and in granivorous rodents that clip fruits directly from trees and herbaceous plants. Postdispersal seed predation arises once seeds have been released from the parent plant. Birds, rodents, and ants are known to be among the most pervasive postdispersal seed predators. Furthermore, postdispersal seed predation can take place at two contrasting stages: predation on the "seed rain" and predation on the "seed bank". Whereas predation on the seed rain occurs when animals prey on released seeds usually flush with the ground surface, predation on the seed bank takes place after seeds have been incorporated deeply into the soil. Both pre- and post-dispersal seed predation are extremely frequent in nature. However, fundamental differences occur between the two. Pre-dispersal predators are most often specialists, as they are more likely to adapt to clustered resources. Predator specialization requires that these animals utilize specific cues like plant chemistry (volatile compounds), color and size to attack seeds. In addition, specialization requires that the life cycle of the predator match the production of seeds by the host plant. Therefore, animals with shorter life span, like insects, are more common pre-dispersal predators. Common insect groups of pre-dispersal predators are Coleoptera, Hemiptera, Hymenoptera and Lepidoptera. For this reason, a greater diversity of predators feed on seeds prior to dispersal. Nevertheless, there are important vertebrate pre-dispersal predators, especially birds and small mammals.

Post-dispersal seed predation is extremely common in virtually all ecosystems. Given the heterogeneity in both resource type (seeds from different species are available), quality (seeds of different ages and/or different status of integrity or decomposition) and location (seeds are scattered and hidden in the environment), most post-dispersal predators have generalist habits. These predators usually belong to a diverse array of animals, such as ants, beetles, crabs, fish, rodents and birds. In addition, not only is the assemblage of post-dispersal seed predators diverse, but its composition varies considerably among ecosystems.

The severity of predation rates also varies between pre- and post-dispersal predation. Although both types of predation can be extremely high, studies simultaneously evaluating both pre- and post-dispersal seed predation suggest that the latter is often more severe than the former. However, seeds already dispersed and not consumed have a chance to germinate, and seeds consumed could die regardless of predation due to constraints imposed by unsuitable germination sites. On the other hand, all non-dispersed seeds predated would be no more than a waste of energy by the parent plant. Therefore, the demographic impacts of pre-dispersal predation on plant populations are likely to be stronger than the impacts of post-dispersal predation.

Effects on Plant Demography

The complex relationship between seed predation and plant demography is an important topic of plant-animal interactive studies. Plant population structure and size over time is closely associated with the effectiveness at which seed predators locate, consume, and disperse seeds. In many cases this relationship is dependent upon the type of seed predator (specialist vs. generalist) or the particular habitat in which the interaction is taking place. The role of seed predation on plant demography may be either detrimental or in particular cases actually beneficial to plant populations.

The Janzen-Connell model is one of the most important studies of how seed predation may influence plant demography. The Janzen-Connell model illustrates how seed density and survival respond to distance from the parent tree and differential rates of seed predation. Seed density is hypothesized to decrease as distance from the parent tree increases. Where seeds are most abundant under the parent tree, seed predation is predicted to be at its highest. As distance from the parent tree increases, seed abundance and thus seed predation are predicted to decrease as seed survival increases.

The degree to which seed predation influences plant populations may vary by whether a plant species is safe site limited or seed limited. If a population is safe site limited it is likely that seed predation will have little impact to the success of the population. In safe site limited populations increased seed abundance does not translate into increased seedling recruitment. However, if a population is seed limited, seed predation has a better chance of negatively affecting the plant population by decreasing seedling recruitment. Maron and Simms found both safe site limited and seed limited populations depending on the habitat in which the seed predation was taking place. In dune habitats seed predators (deer mice) were limiting seedling recruitment in the population, thus negatively affecting the population. However, in grassland habitat the seed predator had little effect on the plant population because it was safe site limited.

In many cases seed predators support plant populations by dispersing seeds away from the parent plant. By dispersing seeds away from the parent plant the seed predator is in effect supporting gene flow between populations. Other seed predators collect seeds and then store or cache them for later consumption. In the case that the seed predator is unable to locate the buried or hidden seed there is a chance that it will later germinate and grow, supporting the species dispersal. Generalist (vertebrate) seed predators may also aid the plant in other indirect ways, for instance by inducing top-down control on host-specific seed predators (termed "intra-guild predation"), and as such negating Janzen-Connell type effects and so benefiting the plant in competition with other plant species.

Palynivore

Honeybee collecting pollen

In zoology, a palynivore is an herbivorous animal which selectively eats the nutrient-rich pollen produced by angiosperms and gymnosperms. Most true palynivores are insects or mites. The category in its strictest application includes essentially all bees, and a few different kinds of wasps, as pollen is often the only solid food consumed by all life stages in these insects (palynivorous mites and thrips typically feed on the liquid content of the pollen grains without actually consuming the exine). The list is expanded greatly if one takes into consideration species where either the larval or adult stage feeds on pollen, but not both; there are other wasps which are in this category, as well as many beetles, flies, butterflies, and moths. There is a vast array of different insects that will feed opportunistically on pollen, as will various birds, orb-weaving spiders and other nectarivores.

Frugivore

A Bornean orangutan (*Pongo pygmaeus*)

A frugivore is a fruit eater. It can be any type of herbivore or omnivorewhere fruit is a preferred food type. Because approximately 20% of all mammalianherbivores also eat fruit, frugivory is common among mammals. Since frugivores eat a lot of fruit, they are highly dependent on the abundance and nutritional composition of fruits. Frugivores can either benefit fruit-producing plants by dispersing seeds, or they can hinder plants by digesting seeds along with the fruits. When both the fruit-producing plant and the frugivore species benefit by fruit-eating behavior, their interaction is called mutualism.

Frugivore Seed Dispersal

Seed dispersal is important for plants because it allows their progeny to move away from their parents over time. The advantages of seed dispersal may have led to the evolution of fleshy fruits, which entice animals to eat the fruits and move the plants seeds from place to place. While many fruit-producing plant species would not disperse far without frugivores, they can usually germinate even if they fall to the ground directly below the parent plant.

Many types of animals are seed dispersers. Mammal and bird species represent the majority of seed-dispersing species. However, frugivorous tortoises, lizards, amphibians, and even fish also disperse seeds. For example, cassowaries are a keystone species because they spread fruit through digestion, and many seeds will not grow unless they have been digested by a cassowary. While frugivores and fruit-producing plant species are present worldwide, there is some evidence that tropical forests have more frugivore seed dispersers than the temperate zone.

Ecological Significance

Frugivore seed dispersal is a common phenomenon in many ecosystems. However, it is not a highly specific type of plant–animal interaction. For example, a single species of frugivorous bird may disperse fruits from several species of plants, or a few species of bird may disperse seeds of one plant species. This lack of specialization could be because fruit availability varies by season and year, which tends to discourage frugivore animals from focusing on just one plant species. Furthermore, different seed dispersers tend to disperse seeds to different habitats, at different abundances, and distances, depending on their behavior and numbers.

Plant Adaptations to Attract Dispersers

There are a number of fruit characteristics that seem to be adaptive characteristics to attract frugivores. Many animal-dispersed fruits advertise their palatability to animals with bright colors and attractive smells (mimetic fruits). Fruit pulp is generally rich in water and carbohydrates and low in protein and lipids. However, the exact nutritional composition of fruits varies widely. The seeds of animal-dispersed fruits are often adapted to survive digestion by frugivores. For example, seeds can become more permeable to water after passage through an animal's gut. This leads to higher germination rates. Some mistletoe seeds even germinate inside the disperser's intestine.

Frugivore Adaptations for Fruit Consumption

In order for frugivores to be good seed dispersers, they must digest fruits without consuming a high proportion of the seeds. Many seed-dispersing animals have specialized digestive systems to process fruits, which leave seeds intact. Some bird species have shorter intestines to rapidly pass seeds from fruits, while some frugivorous bat species have longer intestines. Some seed-dispersing frugivores have short gut-retention times, and others can alter intestinal enzyme composition when eating different types of fruits.

Plant Mechanisms to Delay or Deter Frugivory

Plants invest energy into the production of fruits. Plants have evolved to encourage mutualist frugivores to consume their fruit for seed dispersal, but also evolved mechanisms to decrease consumption of fruits when unripe and from non-seed-dispersing predators. Predators and parasites of fruit include seed predators, insects, and microbial frugivores. Plants have chemical and physical adaptations.

Physical Deterrents:

- Cryptic coloration (e.g. green fruits blend in with the plant leaves)
- Unpalatable textures (e.g. thick skins made of anti-nutritive substances)
- Resins and saps (e.g. prevent animals from swallowing)
- Repellent substances, hard outer coats, spines, thorns

Chemical Deterrents:

- Chemical deterrents in plants are called secondary metabolites. Secondary metabolites are compounds produced by the plant that are not essential for the primary processes, such as growth and reproduction. Toxins might have evolved to prevent consumption by animals that disperse seeds into unsuitable habitats, to prevent too many fruits from being eaten per feeding bout by preventing too many seeds being deposited in one site, or to prevent digestion of the seeds in the gut of the animal. Secondary chemical defenses are divided into three categories: nitrogen-based, carbon-based terpenes, and carbon-based phenolics.

Examples of secondary chemical defenses in fruit:

- Capsaicin is a carbon-based phenolic compound only found in plant genus *Capsicum* (chili and bell peppers). Capsaicin is responsible for the pungent, hot "flavor" of peppers and inhibits growth of microbes and invertebrates.
- Cyanogenic glycosides are nitrogen-based compounds and are found in 130 plant families, but not necessarily in the fruit of all the plants. It is specifically found in the red berries of the genus *Ilex* (holly, an evergreen woody plant). It can inhibit electron transport, cellular respiration, induce vomiting, diarrhea, and mild narcosis in animals.
- Emodin is a carbon-based phenolic compound in plants like rhubarb. Emodin can be cathartic or act as a laxative in humans, kills dipteran larvae, inhibits growth of bacteria and fungi, and deters consumption by birds and mice.
- Starch is a polysaccharide that is slowly converted to fructose as the fruit ripens.

Frugivorous Animals

Birds are a main focus of frugivory research. An article by B.A. Loisell and J.G. Blake, Potential Consequences of Extinction of Frugivorous Birds, discusses the important role frugivorous birds have on ecosystems. The conclusions of their research indicate how the extinction of seed-dispersing species could negatively affect seed removal, seed viability, and plant establishment.

Examples of seed-dispersing birds are the hornbill, the toucan, the aracari, the cotinga (ex. Guianan cock-of-the-rock), and some species of parrots. Frugivores are common in the temperate zone, but mostly found in the tropics. Many frugivorous birds feed mainly on fruits until nesting season, when they incorporate protein-rich insects into their diet. Facultatively-baccivorous birds may also eat bitter berries, such as juniper, in months when alternative foods are scarce.

Mammals are considered frugivorous if the seed is dispersed and able to establish. One example of a mammalian frugivore is the maned wolf, or *Chrysocyon brachyurus*, which is found in South America. A study by J.C. Motta-Junior and K. Martins found that the maned wolf is probably an important seed disperser. The researchers found that 22.5–54.3% of the diet was fruit.

65% of the diet of orangutans consists of fruit. Orangutans primarily eat fruit, along with young leaves, bark, flowers, honey, insects, and vines. One of their preferred foods is the fruit of the durian tree, which tastes somewhat like sweet, cheesy, garlic custard. They discard the skin, eat the flesh, and spit out the seeds.

Other examples of mammalian frugivores include fruit bats and the gray-bellied night monkey, also known as the owl monkey:

> "Owl monkeys are frugivores and supplement their diet with flowers, insects, nectar, and leaves. They prefer small, ripe fruit when available and in order to find these, they forage in large-crown trees (larger than ten meters [32.8 ft]). Seasonal availability of fruit varies across environments. Aotus species in tropical forests eat more fruit throughout the year because it is more readily available compared to the dry forests where fruit is limited in the dry season and owl monkeys are more dependent on leaves."

Conservation

Because seed dispersal allows plant species to disperse to other areas, the loss of frugivores could change plant communities and lead to the local loss of particular plant species. Since frugivore seed dispersal is so important in the tropics, many researchers have studied the loss of frugivores and related it to changed plant population dynamics. Several studies have noted that even the loss of only large frugivores (i.e. monkeys) could have a negative effect, since monkeys are responsible for certain types of long-distance seed dispersal that is not seen with other frugivore types, like birds. However, plant species whose seeds are dispersed by animals may be less vulnerable to fragmentation than other plant species. Frugivores can also benefit from the invasion of exotic fruit-producing species and can be vectors of exotic invasion by dispersing non-native seeds. In short, while anthropogenic habitat loss and change may negatively affect some frugivore species, it may positively affect others.

Nectarivore

In zoology, a nectarivore is an animal which eats the sugar-rich nectar produced by flowering plants. Most nectarivores are incects or birds, but there are also nectarivorous mammals, notably several species of bats in the southwestern United States and Mecico, as well as the Australian Honey Possum (*Tarsipes rostratus*),and Geckos (*genus phesulma*) in Mauritius. The term is less exclusive than other –vores; many animals that are considered *nectarivorous* may also be insectivorous. Many species are nectar robbers, performing no pollination services to a plant while still consuming nectar.

Most nectarivores have relatively long narrow beaks and tongues which seem to be an adaptation for retrieving nectar from flowers. Tongues contain grooves, bristles, and papillae, which increase the surface area and facilitate capillary action to draw up the nectar. Birds can consume the abundant nectar in flowers quickly. A hummingbird can fill up its large crop with nectar and then rest while the crop empties during digestion. Nectarivores tend to have smaller, less muscular gizzards and shorter intestines. The proventricular and pyloric openings of most nectarivores lie in the same plane, so that nectar can bypass the lumen of the gizzard. Ceca are vestigial and the rectum is short.

Nectar is digested rapidly and almost completely. A hummingbird can obtain a metabolizable energy content of glucose, fructose, and sucrose of greater than 95%. To obtain the other nutrients they need, adult hummingbirds spend about 25% of the time foraging for soft-bodied arthropods and increase this effort during breeding and raising young.

Consuming high quantities of nectar imposes the necessity of high rates of excretion and the loss of electrolytes that must be replaced. Such replacement may come primarily from insects, but also from eating sand or licking ashes.

Nectarivores have altricial young with rapid growth and high protein requirements. Adult nectarivores feed their young a predominantly insect diet with increasing amounts of nectar as they age.

Flowers that attract their nectarivore pollinators tend to be red and scentless. The dilute sugar and amino acid concentrations facilitate rapid uptake by the birds, as opposed to higher concentrations in flowers pollinated by insects. Low concentrations may also promote wider foraging by birds and a wider distribution of pollen among the flower population.

Carnivore

A carnivore is an organism that mostly eats meat, or the flesh of animals. Sometimes carnivores are called predators. Organisms that carnivores hunt are called prey.

Carnivores are a major part of the food web, a description of which organisms eat which other organisms in the wild. Organisms in the food web are grouped into trophic, or nutritional, levels. There are three trophic levels. Autotrophs, organisms that produce their own food, are the first trophic level. These include plants and algae. Herbivores, organisms that eat plants and other autotrophs, are the second trophic level. Carnivores are the third trophic level. Omnivores, creatures that consume a wide variety of organisms from plants to animals to fungi, are also the third trophic level.

Autotrophs are called producers, because they produce their own food. Herbivores, carnivores, and omnivores are consumers. Herbivores are primary consumers. Carnivores and omnivores are secondary consumers.

Many carnivores eat herbivores. Some eat omnivores, and some eat other carnivores. Carnivores that consume other carnivores are called tertiary consumers. Killer whales, or orcas, are a classic example of tertiary consumers. Killer whales hunt seals and sea lions. Seals and sea lions are carnivores that consume fish, squid, and octopuses.

Some carnivores, called obligate carnivores, depend only on meat for survival. Their bodies cannot digest plants properly. Plants do not provide enough nutrients for obligate carnivores. All cats, from small house cats to huge tigers, are obligate carnivores.

Most carnivores are not obligate carnivores. A hypercarnivore is an organism that depends on animals for at least 70 percent of its diet. Plants, fungi, and other nutrients make up the rest of their food. All obligate carnivores, including cats, are hypercarnivores. Sea stars, which prey mostly on clams and oysters, are also hypercarnivores.

Mesocarnivores depend on animal meat for at least 50 percent of their diet. Foxes are mesocarnivores. They also eat fruits, vegetables, and fungi.

Hypocarnivores depend on animal meat for less than 30 percent of their diet. Most species of bears are hypocarnivores. They eat meat, fish, berries, nuts, and even the roots and bulbs of plants. Hypocarnivores such as bears are also considered omnivores.

The planet's largest animal is a carnivore. The blue whale can reach 30 meters (100 feet) long and weigh as much as 180 metric tons (200 tons). It feeds by taking huge gulps of water and then filtering out tiny shrimp-like creatures called krill. The blue whale can eat about 3.6 metric tons (4 tons) of krill every day—that's about 40 million of the little creatures. The largest land carnivore is the polar bear, which feeds mainly on seals.

Characteristics of Carnivores

Though carnivores come in many shapes in sizes, but they often have some common similarities. Most carnivores have relatively large brains and high levels of intelligence. They also have less complicated digestive systems than herbivores.

Though all carnivores eat meat at some level, the frequency of their feeding can vary. Warm-blooded carnivores tend to burn a lot of calories. Because of this, they have to hunt and eat often to keep to keep up their energy levels. Cold-blooded carnivores, use less calories and can rest days or even months between meals, for example.

Hunting

Carnivores have biological adaptations that help them hunt. Carnivorous mammals such as wolves have strong jaws and long, sharp teeth that help them grab and rip apart their prey. Plant-eaters, on the other hand, usually have big molars that help them grind up leaves and grasses.

Lions, cougars, and other cats have sharp claws that they use to hunt. Birds such as hawks and owls also hunt with their claws, called talons. Many carnivorous birds, called raptors, have curved beaks that they use to tear apart their prey.

Many carnivores grab their prey in their mouths. Great blue herons wade slowly through shallow water and then suddenly snatch a fish, crab, or other creature from the water. Toads grab mice in their mouths. Sperm whales dive deep into the ocean where they bite hold of squid.

Spiders capture their prey—usually insects—by trapping them in a sticky web. Other carnivores attack their prey with a bite or a sting that injects toxic venom into the victim. The venom either paralyzes or kills the prey. Snakes such as king cobras have hollow fangs that act like needles to inject venom. Cobras mostly prey on other snakes. Jellyfish have stingers on their tentacles, which paralyze fish swimming nearby.

Most carnivores are animals, but plants and fungi can be carnivores also. The Venus flytrap is a plant that catches insects in its leaves. When an insect brushes against the sensitive hairs on the leaf, the leaf folds in two and snaps shut. The insect is trapped inside. Other carnivorous plants, such as the sundew, produce a sticky material that catches insects.

Fungi are a group of organisms that include mushrooms, molds, and mildew. Some fungi trap and consume tiny organisms. Most carnivorous fungi prey on microscopic worms called nematodes, which they trap with suffocating rings.

Diets

Certain types of carnivores have specific diets. Some, such as sea lions, eat mainly fish. They are called piscivores.

Others, such as lizards, eat mainly insects. They are called insectivores. Many bats are also insectivores. One little brown bat can eat a thousand mosquitoes in an hour. Some insects are themselves insectivores. These include ladybugs, dragonflies, and praying mantises.

Carnivores that have been known to attack and eat human beings are known as man-eaters. Some species of sharks, alligators, and bears are called man-eaters. However, no carnivore specifically hunts human beings or relies on them as a regular food source.

Cannibals are carnivores that eat the meat of members of their own species. Many animals practice cannibalism. For some species, cannibalism is a way of eliminating competitors for food, mates, or other resources. Chimpanzees and bears, for example, will hunt and consume the young of family members, sometimes their own offspring. Praying mantis females will kill and eat the bodies of their mates.

Many carnivores are scavengers, creatures that eat the meat of dead animals, or carrion. Unlike other types of carnivores, scavengers usually do not hunt the animals they eat. Some, such as vultures, consume animals that have died from natural causes. Others, such as hyenas, will snatch meat hunted by other carnivores. Many insects, such as flies and beetles, are scavengers.

Some carnivores, including sea lions, feed often. Others, such as king cobras, can go months between meals.

Carnivores in the Food Chain

For a healthy ecosystem, it is important that the populations of autotrophs, herbivores, and carnivores be in balance. Energy from nutrients is lost at each trophic level. It takes many autotrophs to support a fewer number of herbivores. In turn, a single carnivore may have a home range of dozens or even hundreds of miles. A Siberian tiger, for instance, may patrol a range of 1,000 square kilometers (386 square miles).

In some places, the disappearance of large carnivores has led to an overpopulation of herbivores, disrupting the ecosystem. Wolves and cougars are traditional predators of white-tailed deer, for instance. But hunting and development have eliminated these predators from the northeastern United States. Without natural predators, the population of white-tailed deer has skyrocketed. In some areas, there are so many deer that they cannot find enough food. They frequently stray into towns and suburbs in search of food.

Carnivores depend on herbivores and other animals to survive. Zebras and gazelles once traveled in great herds over the plains of Africa. But these herds have shrunk and are now mostly confined to parks and wildlife reserves. As the numbers of these herbivores decline, carnivores such as African wild dogs, which prey on them, also decline. Scientists estimate that only 3,000 to 5,500 African wild dogs remain in the wild.

Piscivore

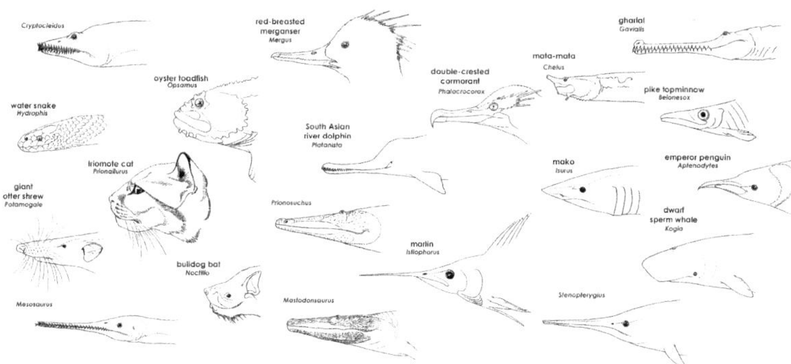

An array of extinct and extant specialist on fish-eating.

Piscivores come in a large array of sizes and morphologies. Not all have teeth. Some are slender-snouted, others broad. What mostly defines a piscivore is that it consumes fish, but not all do it the same way, and for some, it's purely behavioral, sometimes seasonal, and sometimes based on relation to plentiful food.

Some piscivores, including many piscine ones, acquire prey through engulfing their food whole. Indeed, some fish may be classed base on how large a gape they can manage, when it comes to eating the next fish down on the food chain. But this doesn't preclude piscivory in smaller-gaped fish, merely that larger prey cannot be taken. Other piscivores trap their prey in their jaws, using serrated teeth, needle-like teeth, or serrated margins of their jaw, to create high-pressure points along the jaw margin that prevent the slippery prey from wriggling free.

And both of these methods are important, because most piscivores, being largely aquatic by rule, tend not to use any other limb to help restrain prey. From penguins to fish to whales to crocodilians, the jaws are pretty much all you got. This tends to cause piscivores to develop similar adaptations to fish-catching. For one, there's gape, but this more depends on the relationship of prey-size to predator. Another have been the longness or slenderness of the teeth, which may involve either an array of long, piercing teeth or shorter, recurved teeth.

We see the latter in *Hydrophis*, the fish-specialist ocean-going snake, which uses the typical serpentine oral-transport mechanism of "walking prey into the mouth" through the use of flexible joints in the skull. Gharials have long, interlocking teeth, but they mostly orient in the same direction: slightly procumbent, but curved backwards, and when closing the jaws come together so the tips of the teeth are at the same angle. This makes them very efficient "fish-catchers." In contrast, presumed piscivores like the extinct elasmosaurs had teeth that broadly splay out of the jaw and had far less curvature. The jaws of the elasmosaur *Aristonectes* involves much smaller, finer, and less splayed teeth, which would make them comparable to the gharials, despite have a shorter, rounder, and triangular mouth.

One thing that is often brought up in comparison to piscivores is the presence of a radiating row of procumbent teeth at the front of the jaws. However, this, too, varies among aquatic piscivores. The precise mechanism by which procumbent teeth would preclude piscivory is often never explained, though. Instead, it is presumed through analogy, that forward-sloping teeth occur in piscivores, therefore piscivores should have them. This bit of backwards thinking has led to arguments that, for example, some dinosaurs (spinsoaurs, the abelisaur *Masiakasaurus*, *Microraptor*, etc.), pterosaurs (e.g., *Rhamphorhynchus*), and even birds (e.g., *Yanornis)* were piscivores. This "fact" — incidence of forward-sloping teeth relative to diet — has never been qualified across vertebrate taxa. And in fish, it's generally not true: most piscivorous fish either lack teeth (as in marlins) or bear large teeth which are arrayed roughly perpendicular to the jaw margin. Especially in smaller fish, the teeth may even curve backwards, as it does in watersnakes, due to the behavior of oral prehension. In those animals in which food is engaged by the jaws, forward sloping teeth make more inhibitors of prey escape. Instead, rearward sloping or curving teeth, as in most carnivorous *terrestrial* vertebrates, serve to inhibit, as it were, egress from a gaping maw. Even the teeth of fine-food consumers, such as whale and basking sharks, point rearward.

Most of the issues with piscivores in aquatic environments is the segregation of them into two morphotypes: The prehension type, and the engulfment type. Prehensors attempt to restrain prey through use of their jaw shape and/or teeth, whereas engulfers instead attempt to swallow their prey whole. Prehension isn't even restricted to the jaws; fish-eating bulldog bats, ospreys, sea ernes, and flatheaded cats all use claws to catch prey. These predators, however, possess additional limbs; comparison of aquatic oral-only prehensors with terrestrial oral- or limb-based prehensors has yet to receive much scrutiny.

So it's no surprise that for piscivores that have the use of their limbs for catching, subduing, or even processing the carcasses of fish (say, brown bears in spring) they tend not to follow these rules of order for the morphology of the jaws. Why, then, should we assume they do? If teeth that do not work to impale evenly along the jaw, that do not recurve towards the oral margin, that work when biting, make poor teeth for catching prey upon.

Gape is certainly an issue in fish-catching, but it has more to do with relative size of the bits of food going down the gullet. The jaw should not be more capable of holding prey than the throat can swallow; it's sorta the first stop on the way to the stomach. More than one fish has been found having choked to death while trying to swallow prey it barely managed to pass through the mouth, and this is true for snakes as well. If the animal doesn't regurgitate its prey, it will die. Terrestrial animals tend not to involve these factors because many of them develop additional methods of processing prey than mouthing at it with rows of teeth (like many sharks do). They tear at it, pick it apart, then swallow it piece by piece. So while the piscivorous osprey has very recurved talons, it still rips its prey up. Less so the egret, with its relatively small and straight claws, so it swallows prey, and gape is less of an issue that the width of its mandibles, providing space into the gullet. That is also true of the pelican, which will swallow whole other birds that can barely fit into its pouch. At that point, the actual limiting factor is the size of the crop and the width between the shoulders including furculae, passing food from the crop into the stomach.

Examples of Extant Piscivores

- Human
- Wolf
- Aquatic genet
- Flat-headed cat
- Bulldog bat
- Sea lion
- Otter shrew
- European otter
- American mink

- Fishing cat
- Amazon river dolphin
- Giant otter
- Bottlenose dolphin
- Harbor seal
- Osprey
- Merganser
- Penguin
- Bald eagle
- Gharial
- Yellow-bellied sea snake
- African tigerfish
- Barracuda
- Alligator gar
- Lemon shark
- Fishing spider
- Tiger
- Piranha
- Bluefish

Extinct and Prehistoric Piscivores

Specimen of *Diplomystus* swallowing another fish

Numerous extinct and prehistoric animals are hypothesized to have been primarily piscivorous due to anatomy and/or ecology. Furthermore, some have been confirmed to be piscivorous through fossil evidence.

- *Baryonyx* (an opportunistic predator that had a crocodile-like skull, and scales of the lepidotid fish *Scheenstia* have been found in a skeleton where the stomach should be).
- *Spinosaurus* (close relative of *Baryonyx*, is hypothesized to have preyed on fish because of giant coelacanthids found in the same environment, and due to anatomical features, including a pressure-sensitive snout that could have detected movements of swimming prey).
- *Laganosuchus* (flattened head suggests that it passively waited for fish to swim near its mouth in order to engulf them).
- *Pteranodon* (remains of fish found in the beaks and stomach cavities of some specimens).
- *Elasmosaurus* (long neck, stereoscopicly positioned eyes, and long teeth are thought to be adaptations for stalking and trapping fish and other schooling animals).
- *Thyrsocles* (fossil specimen found with the stomach stuffed with the extinct herring *Xyne grex*).
- *Xiphactinus* (a 4-meter-long specimen was found with a perfectly preserved skeleton of its relative, *Gillicus*, in its stomach).
- *Diplomystus* (a small relative of the herring, numerous fossils of individuals that died while trying to swallow other fishes, including smaller individuals of the same species, are known).
- *Ornithocheirus* (hypothesized to be piscivorous due to anatomy of its jaws and dentition).
- *Titanoboa* (multiple cranial and biochemical characteristics suggest it was primarily piscivorous).

Molluscivore

A molluscivore is a carnivorous animal that specialises in feeding on molluscs such as gastropods, bivalves, brachiopods and cephalopods. Known molluscivores include numerous predatory (and often cannibalistic) molluscs, (e.g.octopuses, murexes, decollate snails and oyster drills), arthropods such as crabs and firefly larvae, and, vertebrates such as fish, birds and mammals. Molluscivory is performed in a variety ways with some animals highly adapted to this method of feeding behaviour. A similar behaviour, durophagy, describes the feeding of animals that consume hard-shelled or exoskeleton bearing organisms, such as corals, shelled molluscs, or crabs.

Description

Molluscivory is performed in several ways.

In some cases, the mollusc prey are simply swallowed entire, including the shell, whereupon the prey is killed through suffocation and or exposure to digestive enzymes. Only cannibalistic sea slugs, snail-eating cone shells of the taxon Coninae, and some sea anemones use this method.

One method, used especially by vertebrate molluscivores, is to break the shell, either by exerting force on the shell until it breaks, often by biting the shell, like with oyster crackers, mosasaurs, and placodonts, or hammering at the shell, e.g. oystercatchers and crabs, or by simply dashing the mollusc on a rock (e.g. song thrushes, gulls, and sea otters).

A common glowworm larva hunting snails

Another method is to remove the shell from the prey. Molluscs are attached to their shell by strong muscular ligaments, making the shell's removal difficult. Molluscivorous birds, such as oystercatchers and the Everglades snail kite, insert their elongate beak into the shell to sever these attachment ligaments, facilitating removal of the prey. The carnivorous terrestrial pulmonate snail known as the "decollate snail" ("decollate" being a synonym for "decapitate") uses a similar method: it reaches into the opening of the prey's shell and bites through the muscles in the prey's neck, whereupon it immediately begins devouring the fleshy parts of its victim. The walrus sucks meat out of bivalve molluscs by sealing its powerful lips to the organism and withdrawing its piston-like tongue rapidly into its mouth, creating a vacuum.

Another method is used by many molluscs, themselves. Octopi, nautilii, and most molluscivoruous snails use their radula to drill a hole through the shell, then inject venom and digestive enzymes through the hole, whereupon the digested prey is then sucked out through the hole.

The larvae of glowworms and fireflies are simply small enough to enter the shells of terrestrial snails and begin eating immediately.

In Marine Mammals

The walrus, an aquatic molluscivore

Whales: Sperm whales, pilot whales, Cuvier's beaked whale, Risso's dolphin and species in the genera *Mesoplodon*, and *Hyperoodon* and the superfamily Physeteroidea are classified as molluscivores, eating mainly squid.

Pinnipeds: Elephant seals, Ross seals and South American fur seals are classed as molluscivores. The walrus eats benthic bivalve molluscs, especially clams, for which it forages by grazing along the sea bottom, searching and identifying prey with its sensitive vibrissae. The walrus sucks the meat out by sealing its powerful lips to the organism and withdrawing its piston-like tongue rapidly into its mouth, creating a vacuum. The walrus palate is uniquely vaulted, enabling effective suction.

In Fish

The spotted pufferfish, a molluscivore

Several species of pufferfish and loaches are molluscivores. As many molluscs are protected by a shell, the feeding techniques applied amongst molluscivore fish are highly specialized and usually divided into two groups: "crushers" and "slurpers." Pufferfish tend to be crushers and will use their beak-like teeth to break the shell in order to gain access to the meat inside. Loaches are specialized slurpers, and will make use of their characteristically shaped snout in order to grab hold of the animal living inside the shell.

The black carp (*Mylopharyngodon piceus*) commonly feeds by crushing large molluscs with pharyngeal teeth, extracting soft tissue, and spitting out shell fragments. Four-year-old juveniles are capable of consuming approximately 1–2 kg of molluscs per day. This bottom-dwelling molluscivore was purposely imported into the United States in the early 1970s for use as a food fish and also as a biological control agent for snails—an intermediate host for a trematode parasite in fish reared on fish farms. Two snail-eating cichlids, *Trematocranus placodon* and *Maravichromis anaphyrmis*, have been tried as biological control agents of schistosomes in fish ponds in Africa. Redear sunfish (*Lepomis micropholus*) and bluegill (*Lepomis macrochirus*) have been used to control quagga mussels (*Dreissena bugensis*) in the lower Colorado River in the US.

The common name of some fish reflects their molluscivorous feeding, for example, the "snail-crusher hap" (*Trematocranus placodon*), """red rock sheller" (*Haplochromis sp.*), "Rusinga oral sheller" (*Haplochromis sp.*) and "rainbow sheller" (*Haplochromis sp.*). The redear sunfish is also known as the "shellcracker".

In Reptiles

Gray's monitor (or "butaan") is well known for its diet, which consists primarily of ripe fruit; however, several prey items are also consumed, including snails. Monitors are generally carnivorous animals, which makes the Gray's monitor somewhat of an exception amongst the varanid family.

The prehistoric placodont reptiles is an extinct taxon of marine animals that superficially resembled lizards and turtles, most of whose dentition of peg-like incisors and enormous, molar-like teeth allowed them to prey on molluscs and brachiopods by plucking their prey off of the substrate, and crushing the shells.

In Birds

Among birds, the eponymous shorebirds known as oystercatchers are renowned for feeding upon bivalves. At least one bird of prey is also primarily a molluscivore—the snail kite, *Rostrhamus sociabilis*. The limpkin is a small rail-like bird that feeds almost entirely on apple snails. Other birds that will eat molluscs occasionally include mergansers, ducks, coots, dippers and spoonbills.

In Invertebrates

Cone snail feeding on a cowrie

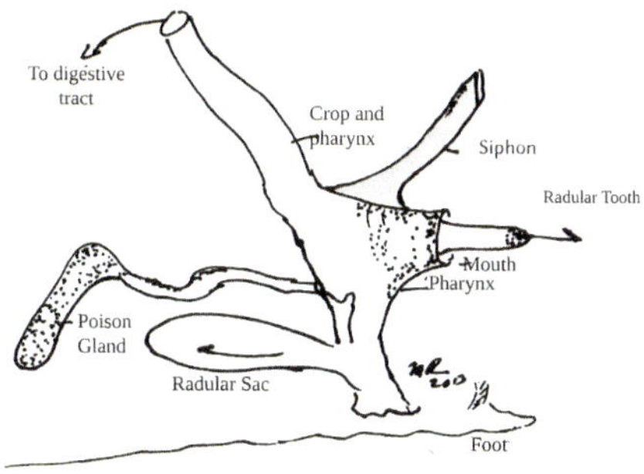

Venom apparatus of a cone snail

Cone snails: Some cone snails hunt and eat other kinds of snails, such as cowries, olive shells, turbo snails, and conch snails, while others will eat other cone snails. *Conus marmoreus* and *Conus omaria* are able to kill and swallow prey that are larger than themselves; some Conus species can swallow prey that weigh up to half of their own weight. Snail's bodies are attached to their shell by a columellar muscle that holds onto the columella, the axis of the snail. This muscle also allows the snail to retract back into its shell. If this muscle is broken, the snail will lose its shell and die. It is hard to detach this muscle in a live snail, or even in a dead snail. It is thought that the conotoxins in the venom of cone snails are able to completely relax this muscle so that the body can be pulled out from its shell. The cone snail uses its foot to hold the shell of its prey. Using a strong, steady pulling motion, the body of the snail can be forced out and then swallowed whole. Complete digestion of a snail can take many hours, even days.

Starfish: Primitive starfish, such as *Astropecten* and *Luidia*, swallow their prey whole and start to digest it in their cardiac stomachs. Shell valves and other inedible materials are ejected through their mouths. The semi-digested fluid is passed into their pyloric stomachs and caeca where digestion continues and absorption occurs. The margined sea star (*Astropecten articulatus*) is a well known molluscivore. It catches prey with its arms which it then takes to the mouth. The prey is then trapped by the long, moving prickles around the mouth cavity and swallowed food.

In more advanced species of starfish, the cardiac stomach can be everted from the organism's body to engulf and digest food. When the prey is a clam, the starfish pulls with its tube feet to separate the two valves slightly, and inserts a small section of its stomach, which releases enzymes to digest the prey. The stomach and the partially digested prey are later retracted into the disc. Here the food is passed on to the pyloric stomach, which always remains inside the disc. Because of this ability to digest food outside the body, starfish can hunt prey much larger than their mouths.

Crabs: The freshwater crabs *Syntripsa matannensis* and *Syntripsa flavichela* are classed as molluscivores. Using their massive and powerful claws, adult Florida stone crabs (*Menippe mercenaria*) feed on acorn barnacles, hard-shelled clams, scallops, and conch.

Vermivore

Pacific brown salamander eating a worm

Vermivore is a zoological term for animals that eat worms (including annelids, nematodes, and other worm-like animals). Animals with such a diet are known to be vermivorous. Some definitions are less exclusive with respect to the diet, but limit the definition to particular animals, e.g. "Feeding on worms or insect vermin. Used of a bird."

An entire genus of New World warblers has been given the name *Vermivora*.

One vermivore that may feed exclusively on worms is *Paucidentomys vermidax*, a rodent species of a type commonly known as shrew rats which was discovered in 2011 in Indonesia. The name, which can be translated as "worm-eating, few-toothed mouse", refers to the fact that they have only four teeth and may live exclusively on a diet of earthworms. This reduced dentition in vermivorous mammals is said to be due to relaxed selectional pressure on dental occlusion.

Examples

- Shrews
- Tuataras
- Hedgehogs
- Kiwi
- Long-beaked echidna
- Platypus
- Leeches of the genus *Americobdella*
- Moles
- American robins
- Wormsnakes
- Some sea snails e.g. *Jaspidiconus* and *Conus* species.

Insectivore

Insectivore, the common name applied to any of 450 or so species of mammals—comprising hedgehogs, golden moles, "true" moles, "true" shrews, the moonrat, gymnures, solenodons, and tenrecs—that subsist primarily on insects, other arthropods, and earthworms.

Insectivora is obsolete as a taxonomic order, but the term *insectivore* is still used to refer to the remaining members, which have been classified into three orders: Soricimorpha, Erinaceomorpha, and Chrysochloridea. Together these three orders are called grandorder Lipotyphla by mammalogists, its members being referred to as either lipotyphlans or insectivores.

Western European hedgehog (*Erinaceus europaeus*).

Classification

Classification of insectivores has been problematic since the founding of modern taxonomy by Carolus Linnaeus. In 1758 he placed hedgehogs, shrews, and moles with opossums, armadillos, and pigs in the order Bestiae, as all have a long snout. In 1795 hedgehogs, shrews, and moles were separated from the others and joined with bears and other carnivores in the order Plantigrades, because all walk on the soles of their feet (plantigrade). Hedgehogs, shrews, and moles were placed in the family Subterranea, order Faculata, in 1811, but all had been recognized as "insectivores" by 1816, and that common name was latinized as Insectivora in 1821. By 1855 the first few genera of tree shrews and elephant shrews had been described and included within Insectivora, and in 1866 the order was divided into suborder Menotyphla, containing tree shrews and elephant shrews, and suborder Lipotyphla for hedgehogs, shrews, moles, and their relatives. Revised classifications followed, but owing to the wide range of anatomical features present in the group, specialists could agree only that insectivores were placental mammals that could not be placed in any of the better-defined orders. Insectivora thus became basically a taxonomic "catch-all" group containing those mammals impossible to place elsewhere.

Common Eurasian shrew (*Sorex araneus*).

golden-rumped elephant shrewLearn about the golden-rumped elephant shrew (*Rhynchocyon chrysopygus*).

The definition of Insectivora was significantly modified in 1956 by the insectivore specialist P.M. Butler, who raised Lipotyphla to the rank of order (now containing only hedgehogs, shrews, moles, solenodons, tenrecs, and golden moles), assigned elephant shrews and tree shrews to separate orders, and abandoned the name Insectivora. Most professional mammalogists now use Lipotyphla instead of Insectivora and informally refer to the group as either lipotyphlans or insectivores. Others, however, prefer Insectivora, and that name is still used in lay publications as well as some professional compilations.

Defining the Lipotyphla has remained intractable, however, and the closest living relative of Lipotyphla has not been determined. Living lipotyphlans apparently share only two traits for certain. All lipotyphlans lack a cecum and have a short gut tube (large and small intestine) relative to body length without any expansion or shape change toward the end. In addition, the cartilaginous connection between the pubic bones of the pelvis is reduced. Most lipotyphlans also have a large extension of the maxillary bone that forms part of the eye socket. The most authoritative classification of mammals, presented by M.C. McKenna and Susan K. Bell in 1997, placed Lipotyphla at a taxonomic rank higher than order (grandorder) and recognized three orders within the grandorder: Chrysochloridea (golden moles), Erinaceomorpha (hedgehogs, gymnures, the moonrat, and moles), and Soricimorpha (solenodons, shrews, and tenrecs). The following arrangement basically follows the classification by McKenna and Bell (1997), with a few alterations.

Grandorder Lipotyphla (Insectivores)

450 or so species in 64 genera and 3 orders. Fossils of 171 extinct genera and 13 extinct families have been identified. The earliest lipotyphlan fossils are from the order Soricimorpha and date to the Late Cretaceous Period in North America and Asia.

- Order Chrysochloridea
    - Family Chrysochloridae (golden moles)

      18 species in 7 genera from Africa. Tenrecs (family Tenricidae) may actually be a member of this group.
    - Order Erinaceomorpha

      23 species in 1 family. 4 fossil families contain 15 genera and 9 additional genera unallocated to families dating to the Early Paleocene Epoch. The group's evolutionary relationship with other lipotyphlans and even with other mammalian orders is unresolved. Moles (family Talpidae) are sometimes classified in this group.
        - Family Erinaceidae (hedgehogs, gymnures, and the moonrat)

          23 species in 7 genera. 22 fossil genera dating to the Paleocene in North America, the Eocene in Europe and Asia, and the Miocene in Africa.

        - Subfamily Erinaceinae (hedgehogs)

          15 species in 4 genera from Europe, Asia, and Africa.

        - Subfamily Galericinae (moonrat and gymnures)

          8 species in 3 genera from Southeast Asia.

- Order Soricimorpha

  More than 400 species in 4 families. 9 fossil families contain 30 genera, some dating to the Late Cretaceous. Moles (family Talpidae) are sometimes classified with hedgehogs in Erinaceomorpha.

    - Family Soricidae (true shrews)

      341 or more species in 23 genera and 60 extinct genera dating to the Middle Eocene in North America, the Late Eocene in Europe, the Early Oligocene in Asia, the Middle Miocene in Africa, and the Pleistocene in northwestern South America.

        - Subfamily Crocidurinae

          208 species in 9 genera.

        - Subfamily Soricinae

          116 or more species in 11 genera.

        - Subfamily Myosoricinae

          17 species in 3 genera.

    - Family Talpidae (true moles)

      42 species in 17 genera. 30 fossil genera have been identified from the Middle Eocene of Europe, the Oligocene in Asia and the Mediterranean region, and the Late Oligocene in North America.

        - Subfamily Talpinae

          36 species in 14 genera.

        - Subfamily Uropsilinae (Asiatic shrew moles)

          4 species in 1 genus.

        - Subfamily Desmaninae (desmans)

          2 species in 2 genera.

- Family Tenrecidae(tenrecs)

  29 species in 10 genera. 3 fossil genera dating to the Early Miocene in Africa, to the Pleistocene on Madagascar.

  - Subfamily Oryzorictinae

    21 species in 3 genera.

  - Subfamily Tenrecidae

    4 species in 4 genera.

  - Subfamily Potamogalinae (otter shrews)

    3 species in 2 genera.

  - Subfamily Geogalinae (large-eared tenrec)

    1 species.

- Family Solenodontidae(solenodons)

  2 species in 1 genus.

Fungivore

Fungivory is feeding on mycelia, fruit bodies, or spores of fungi. Obligate or facultative fungivory is characteristic of many saproxylic (dead wood and coarse woody debris dependent) insects, which exploit the lignolytic and cellulolytic abilities of fungi to feed on mycelia and inhabit wood altered by fungi. At least six types of fungivory can be identified based on the life stages of the insect (larva and adult) and the fungus (spore, mycelium, and fruit body). Furthermore, fungivorous insects can be divided into consumers of living and of dead fungi. Lawrence separates micro- and macrophagy of fungivores based on the mouthpart morphology of beetles and the types of food sources. All types of trophic specialization can be found among the fungivores from mono- and oligophagy to the most common polyphagy – in an extreme case, an Australasian fungivore Cis bilamellatus invaded northern Europe using a broad range of fungal hosts. Associations between fungi and insects are among the most species-rich, diverse, and complex in terrestrial ecosystems. The unique adaptations and dispersal abilities of both fungi and insects result in a variety of interaction types. One of the main guilds is comprised of fungivorous beetles. The high species diversity of fungivores in distantly related phylogenetic clades of Coleoptera indicates a special role of fungi in the evolution of the largest order of living organisms. Fungivory in Coleoptera seems to have evolved independently several times. The paleontological evidence of insect fungivory is scarce – large fruit bodies of the Basidiomycetes are unknown until the Early Cretaceous, but a recent discovery from amber in France suggests that macromycetes have served as a habitat for fungivorous insects since the Albian, ca 100 million years ago. Although insects often consume mycelia together with their woody substrates, the focus of this review is on beetles selectively consuming fruit bodies and spores of fungi. Most of the publications on insect fungivory have dealt with fruit bodies, and the present review aims to summarise these studies and to identify future research directions.

Animals

Mammals

Many mammals eat fungi, but only a few feed exclusively on fungi; most are opportunistic feeders and fungi only make up part of their diet. At least 22 species of primate, including humans, bonobos, colobines, gorillas, lemurs, macaques, mangabeys, marmosets and vervet monkeys are known to feed on fungi. Most of these species spend less than 5% of the time they spend feeding eating fungi, and fungi therefore form only a small part of their diet. Some species spend longer foraging for fungi, and fungi account for a greater part of their diet; buffy-tufted marmosets spend up to 12% of their time consuming sporocarps, Goeldi's monkeys spend up to 63% of their time doing so and the Yunnan snub-nosed monkey spends up to 95% of its feeding time eating lichens. Fungi are comparatively very rare in tropical rainforests compared to other food sources such as fruit and leaves and they are also distributed more sparsely and appear unpredictably, making them a challenging source of food for Goeldi's monkeys.

Fungi are renowned for their poisons to deter animals from feeding on them: even today humans die from eating poisonous fungi. A natural consequence of this is the virtual absence of obligate vertebrate fungivores, with the diprotodont family Potoridae being the major exception. One of the few extant vertebrate fungivores is the northern flying squirrel, but it is believed that in the past there were numerous vertebrate fungivores and that toxin development greatly lessened their number and forced these species to abandon fungi or diversify. Although some monkeys still eat fungi today, there are no completely fungivorous primates, though their dentition is very suitable for eating fungi.

Mollusks

Many terrestrial gastropod mollusks are known to feed on fungi. It is the case in several species of slugs from distinct families. Among them are the Philomycidae (e. g. *Philomycus carolinianus* and *Phylomicus flexuolaris*) and *Ariolimacidae* (*Ariolimax californianus*), which respectively feed on slime molds (myxomycetes) and mushrooms (basidiomycetes). Species of mushroom producing fungi used as food source by slugs include milk-caps, *Lactarius* spp., the oyster mushroom, *Pleurotus ostreatus* and the penny bun, *Boletus edulis*. Other species pertaining to different genera, such as *Agaricus*, *Pleurocybella* and *Russula*, are also eaten by slugs. Slime molds used as food source by slugs include *Stemonitis axifera* and *Symphytocarpus flaccidus*. Some slugs are selective towards certain parts or developmental stages of the fungi they eat, though this behavior varies greatly. Depending on the species and other factors, slugs eat only fungi at specific stages of development. Moreover, in other cases, whole mushrooms can be eaten, without any trace of selectivity.

A banana slug feeding on *Amanita*

Insects

Euprenolepis procera, the only species of ant known to harvest mushrooms, feeding on a *Pleurotus* mushroom

In 2008, *Euprenolepis procera* a species of ant from the rainforests of South East Asia was found to harvest mushrooms from the rainforest. Witte & Maschwitz found that their diet consisted almost entirely of mushrooms, representing a previously undiscovered feeding strategy in ants. Several beetle families, including the Erotylidae, Endomychidae, and certain Tenebrionidae also are specialists on fungi, though they may eat other foods occasionally. Other insects, like fungus gnats and scuttle flies, utilize fungi at their larval stage. Feeding on fungi is crucial for dead wood eaters as this is the only way to acquire nutrients not available in nutritionally scarce dead wood.

Birds

Jays (*Perisoreus*) are believed to be the first birds in which mycophagy was recorded. Canada jays (*P. canadensis*), Siberian jays (*P. infaustus*) and Oregon jays (*P. obscurus*) have all been recorded to eat mushrooms, with the stomachs of Siberian jays containing mostly fungi in the early winter. The ascomycete, *Phaeangium lefebvrei* found in north Africa and the Middle East is eaten by migrating birds in winter and early spring, mainly be species of lark (Alaudidae). Bedouin hunters have been reported to use *P. lefebvrei* as bait in traps to attract birds.

Fungi are known to form an important part of the diet of the southern cassowary (*Casuarius casuarius*) of Australia. Bracket fungi have been found in their droppings throughout the year, and Simpson in the *Australasian Mycological Newsletter* suggested it is likely they also eat species of Agaricales and Pezizales but these have not been found in their droppings since they disintegrate when they are eaten. Emus (*Dromaius novaehollandiae*) will eat immature *Lycoperdon* and *Bovista* fungi if presented to them as will brush turkeys (*Alectura lathami*) if offered *Mycena*, suggesting that species of Megapodiidae may feed opportunistically on mushrooms.

Microbial

Fungi

Mycoparasitism occurs when any fungus feeds on other fungi, a form of parasitism, our knowledge of it in natural environments is very limited. *Collybia* grow on dead mushrooms.

The fungal genus, *Trichoderma* produces enzymes such as chitinases which degrade the cell walls of other fungi. They are able to detect other fungi and grow towards them, they then bind to the hyphae of other fungi using lectins on the host fungi as a receptor, forming an appressorium. Once this is formed, *Trichoderma* inject toxic enzymes into the host and probably peptaibol antibiotics, which create holes in the cell wall, allowing *Trichoderma* to grow inside of the host and feed. *Trichoderma* are able to digest sclerotia, durable structures which contain food reserves, which is important if they are to control pathogenic fungi in the long term. *Trichoderma* species have been recorded as protecting crops from *Botrytis cinerea*, *Rhizoctonia solani*, *Alternaria solani*, *Glomerella graminicola*, *Phytophthora capsici*, *Magnaporthe grisea* and *Colletotrichum lindemuthianum*; although this protection may not be entirely due to *Trichoderma* digesting these fungi, but by them improving plant disease resistance indirectly.

Bacteria

Bacterial mycophagy was a term coined in 2005, to describe the ability of some bacteria to "grow at the expense of living fungal hyphae". In a 2007 review in the *New Phytologist* this definition was adapted to only include bacteria which play an active role in gaining nutrition from fungi, excluding those that feed off passive secretions by fungi, or off dead or damaged hyphae. The majority of our knowledge in this area relates to interactions between bacteria and fungi in the soil and in or around plants, little is known about interactions in marine and freshwater habitats, or those occurring on or inside animals. It is not known what affects bacterial mycophagy has on the fungal communities in nature.

There are three mechanisms by which bacteria feed on fungi; they either kill fungal cells, cause them to secrete more material out of their cells or enter into the cells to feed internally and they are categorised according to these habits. Those that kill fungal cells are called nectrotrophs, the molecular mechanisms of this feeding are thought to overlap considerably with bacteria that feed on fungi after they have died naturally. Necrotrophs may kill the fungi through digesting their cell wall or by producing toxins which kill fungi, such as tolaasin produced by *Pseudomonas tolaasii*. Both of these mechanisms may be required since fungal cell walls are highly complex, so require many different enzymes to degarde them, and because experiments demonstrate that bacteria that produce toxins cannot always infect fungi. It is likely that these two systems act synergistically, with the toxins killing or inhibiting the fungi and exoenzymes degrading the cell wall and digesting the fungus. Examples of necrotrophs include *Staphylococcus aureus* which feed on *Cryptococcus neoformans*, *Aeromonas caviae* which feed on *Rhizoctonia solani*, *Sclerotium rolfsii* and *Fusarium oxysporum*, and some myxobacteria which feed on *Cochliobolus miyabeanus* and *Rhizoctonia solani*.

Bacteria which manipulate fungi to produce more secretions which they in turn feed off are called extracellular biotrophs; many bacteria feed on fungal secretions, but do not interact directly with the fungi and these are called saprotrophs, rather than biotrophs. Extracellular biotrophs could alter fungal physiology in three ways; they alter their development, the permeability of their membranes (including the efflux of nutrients) and their metabolism. The precise signalling molecules that are used to achieve these changes are unknown, but it has been suggested that auxins (better known for their role as a plant hormone) and quorum sensing molecules may be involved. Bacteria have been identified that manipulate fungi in these ways, for example mycorhizzal helper bacteria (MHBs) and *Pseudomonas putida*, but it remains to be demonstrated whether the changes they make are cause are directly beneficial to the bacteria. In the case of MHBs, which increase infection of plant roots by mycorrhizal fungi, they may benefit, because the fungi gain nutrition from the plant and in turn the fungi will secrete more sugars.

The third group, that enter into living fungal cells are called endocellular biotrophs. Some of these are transmitted vertically whereas others are able to actively invade and subvert fungal cells. The molecular interactions involved in these interactions are mostly unknown. Many endocellular biotrophs, for example some *Burkholderia* species, belong to the β-proteobacteria which also contains species which live inside the cells of mammals and amoeba. Some of them, for example *Candidatus* Glomeribacter gigasporarum, which colonises the spores of *Gigaspora margarita*, have reduced genome sizes indicating that they have become entirely dependent on the metabolic functions of the fungal cells in which they live. When all the endocellular bacteria inside *G. margarita* were removed, the fungus grew differently and was less fit, suggesting that some bacteria may also provide services to the fungi they live in.

Ciliates

The ciliate family Grossglockneridae, including the species *Grossglockneria acuta*, feed exclusively on fungi. *G. acuta* first attaches themselves to a hyphae or sporangium via a feeding tube and then a ring-shaped structure, around 2 μm in diameter is observed to appear on the fungus, possibly consisting of degraded cell wall material. *G. acuta* then feeds through the hole in the cell wall for, on average, 10 minutes, before detaching itself and moving away. The precise mechanism of feeding is not known, but it conceivably involves enzymes including acid phosphatases, cellulases and chitinases. Microtubules are visible in the feeding tube, as are possible reserves of cell membrane, which may be used to form food vacuoles filled with the cytoplasm of the fungus, via endocytosis, which are then transported back into *G. acuta*. The holes made by *G. acuta* bear some similarities to those made by amoeba, but unlike amoeba *G. acuta* never engulfs the fungus.

Plants

Monotropastrum humile, a myco-heterotroph dependent on fungi throughout its lifetime

Around 90% of land plants live in symbiosis with mycorrhizal fungi, where fungi gain sugars from plants and plants gain nutrients from the soil via the fungi. Some species of plant have evolved to manipulate this symbiosis, so that they no longer give fungi sugars that they produce and instead gain sugars from the fungi, a process called myco-heterotrophy. Some plants are only dependent on fungi as a source of sugars during the early stages of their development, these include most of the orchids as well as many ferns and lycopods. Others are dependent on this food source for their entire lifetime, including some orchids and Gentianaceae, and all species of Monotropaceae and Triuridaceae. Those that are dependent on fungi, but still photosynthesise are called mixotrophs since they gain nutrition in more than one way, by gaining a significant amount of sugars from fungi, they are able to grow in the deep shade of forests. Examples include the orchids *Epipactis*, *Cephalanthera* and *Plantanthera* and the tribe Pyroleae of the family Ericaceae. Others, such as *Monotropastrum humile*, no longer photosynthesise and are totally dependent on fungi for nutrients. Around 230 such species exist, and this trait is thought to have evolved independently on five occasions outside of the orchid family. Some individuals of the orchid species *Cephalanthera damasonium* are mixotrophs, but others do not photosynthesise. Because the fungi that myco-heterotrophic plants gain sugars from in turn gain them from plants that do photosynthesise, they are considered indirect parasites of other plants. The relationship between orchids and orchid mycorrhizae has been suggested to be somewhere between predation and parasitism.

The precise mechanisms by which these plants gain sugars from fungi are not known and has not been demonstrated scientifically. Two pathways have been proposed; they may either degrade fungal biomass, particularly the fungal hyphae which penetrate plant cells in a similar manner to in arbuscular mycorrhizae, or absorb sugars from the fungi by disrupting their cell membranes, through mass flow. To prevent the sugars returning to the fungi, they must compartmentalise the sugars or convert them into forms which the fungi cannot use.

Fungal Farming

Insects

Three insect lineages, the beetles, ants and termites, independently evolved the ability to farm fungi between 40 and 60 million years ago. In a similar way to the way that human societies became more complex after the development of plant-based agriculture, the same occurred in these insect lineages when they evolved this ability and these insects are now of major importance in ecosystems. The methods that insects use to farm fungi share fundamental similarities with human agriculture. Firstly, insects inoculate a particular habitat or substrate with fungi, much in the same as humans plant seeds in fields. Secondly, they cultivate the fungi by regulating the growing environment to try to improve the growth of the fungus, as well as protecting it from pests and diseases. Thirdly they harvest the fungus when it is mature and feed on it. Lastly they are dependent on the fungi they grow, in the same way that humans are dependent on crops.

Beetles

Gallery of *Xylosandrus crassiusculus* split open, with larvae and black fungus

Ambrosia beetles, for example *Austroplatypus incompertus*, farm ambrosia fungi inside of trees and feed on them. The mycangia (organs which carry fungal spores) of ambrosia beetles contain various species of fungus, including species of *Ambrosiomyces*, *Ambrosiella*, *Ascoidea*, *Ceratocystis*, *Dipodascus*, *Diplodia*, *Endomycopsis*, *Monacrosporium* and *Tuberculariella*. The ambrosia fungi are only found in the beetles and their galleries, suggesting that they and the beetles have an obligate symbiosis.

Termites

Termitomyces mushrooms growing out of a termite nest

Around 330 species of termites in twelve genera of the subfamily Macrotermitinae cultivate a specialised fungus in the genus *Termitomyces*. The fungus is kept in a specialised part of the nest in fungus cones. Worker termites eat plant matter, producing faecal pellets which they continuously place on top of the cone. The fungus grows into this material and soon produces immature mushrooms, a rich source of protein, sugars and enzymes, which the worker termites eat. The nodules also contain indigestible asexual spores, meaning that the faecal pellets produced by the workers always contain spores of the fungus that colonise the plant material that they defaecate. The *Termitomyces* also fruits, forming mushrooms above ground, which mature at the same time that the first workers emerge from newly formed nests. The mushrooms produce spores that are wind dispersed, and through this method, new colonies acquire a fungal strain. In some species, the genetic variation of the fungus is very low, suggesting that spores of the fungus are transmitted vertically from nest to nest, rather than from wind dispersed spores.

Ants

Around 220 described species, and more undescribed species of ants in the tribe Attini cultivate fungi. They are only found in the New World and are thought to have evolved in the Amazon Rainforest, where they are most diverse today. For these ants, farmed fungi are the only source of food on which their larvae are raised on and are also an important food for adults. Queen ants carry a small part of fungus in small pouches in their mouthparts when they leave the nest to mate, allowing them to establish a new fungus garden when they form a new nest. Different lineages cultivate fungi on different substrates, those that evolved earlier do so on a wide range of plant matter, whereas leaf cutter ants are more selective, mainly using only fresh leaves and flowers. The fungi are members of the Lepiotaceae and Pterulaceae families. Other fungi in the genus *Escovopsis* parasitise the gardens and antibiotic-producing bacteria also inhabit the gardens.

Humans

Gastropods

The marine snail *Littoraria irrorata*, which lives in the salt marshes of the southeast of the United States feeds on fungi that it encourages to grow. It creates and maintains wounds on the grass, *Spartina alterniflora* which are then infected by fungi, probably of the *Phaeosphaeria* and *Mycosphaerella* genera, which are the preferred diet of the snail. They also deposit faeces on the wounds that they create, which encourage the growth of the fungi because they are rich in nitrogen and fungal hyphae. Juvenile snails raised on uninfected leaves do not grow and are more likely to die, indicating the importance of the fungi in the diet of *L. irrorata*.

Omnivore

An omnivore is an organism that regularly consumes a variety of material, including plants, animals, algae, and fungi. They range in size from tiny insects like ants to large creatures—like people.

Human beings are omnivores. People eat plants, such as vegetables and fruits. We eat animals, cooked as meat or used for products like milk or eggs. We eat fungi such as mushrooms. We also eat algae, in the form of edible seaweeds such as nori, which are used to wrap sushi rolls, and sea lettuce, eaten in salads. Bears are omnivores, too. They eat plants like berries as well as mushroom fungi and animals like salmon or deer

Omnivores are a major part of the food web, a description of which organisms eat which other organisms in the wild. Organisms in the food web are grouped into trophic, or nutritional, levels. There are three trophic levels. Autotrophs, organisms that produce their own food, are the first trophic level. These include plants and algae. Herbivores, organisms that consume plants and other autotrophs, are the second trophic level. Both omnivores and carnivores, meat eaters, are the third trophic level

Autotrophs are called producers, because they produce their own food. Herbivores, carnivores, and omnivores are consumers. Herbivores are primary consumers. Carnivores and omnivores are secondary consumers.

Most birds are omnivores. Robins pull worms from the ground. They also feast on berries. Ostriches graze on plants and grasses. They also eat lizards and insects.

Many mammals are omnivorous. Skunks eat rodents, lizards, honeybees, leaves, grasses, nuts, fungi, and almost anything else they can find.

Some reptiles are also omnivorous. Box turtles feed on fish, frogs, rodents, and many other creatures, but they also eat flowers, berries, and roots.

Fish can also be omnivorous. The opaleye, a fish that feeds mostly on seaweeds along the Pacific Coast of North America, also eats small creatures found among the seaweed.

Some insects are omnivores. Ants eat seeds, nectar, and, often, other insects.

Some omnivores are scavengers, creatures that eat the meat of dead animals. Black bears eat mostly nuts, berries, and other fruit. But if they find a dead animal, they eat it.

Many animals that are often thought of as carnivores are in fact omnivores. Red foxes, for example, prey on rabbits, but they also eat fruit.

Some animals that are thought of as herbivores also eat animals. Squirrels eat mostly nuts, fruits, and seeds, but they sometimes eat insects, small birds, and other creatures.

Omnivore Adaptations

Many omnivores have biological adaptations that help them eat a variety of kinds of foods. They have adapted many characteristics of both carnivores and herbivores. Like many carnivores, raccoons have sharp front teeth that help them rip apart mice and other small creatures. And like many herbivores, raccoons also have large molars that help them chew up plants. Raccoons also have quick paws and long fingers that they can use both to grab prey and to reach a variety of fruits and other plant products.

Compared to herbivores and carnivores, omnivores often have a greater chance of surviving difficult conditions. They can adjust their diets. If all the salmon or other animals disappear from a river ecosystem, a big cat living in that habitat could not survive. Cats are carnivores that cannot digest or obtain nutrients from plant material. However, a grizzly bear could still survive eating berries, fruit, roots, and insects.

Because they have an easier time finding food, omnivores are sometimes better at adapting to new environments than creatures with more specific feeding habits. Omnivores can better adapt to development than herbivores or carnivores.

Urban development, the process of clearing land for homes, business, and agriculture, destroys habitats, the places where animals live in the wild. Herbivores such as elephants cannot survive without a lot of trees and grasses to eat. But omnivores such as opossums, seagulls, and many species of monkey easily adapt to life in urban areas and farmland, where they often find meals in garbage cans.

Size

Animal omnivores (including humans) come in many different sizes. The largest terrestrial omnivore is the endangered Kodiak bear. It can grow up to 10 feet tall (3.04 meters) and weigh up to 1,500 lbs. (680 kilograms). Kodiaks eat grass, plants, fish, berries and the occasional mammal.

Ants may be the smallest omnivores. One of the smallest ants is the pharaoh ant, which grows to only 0.04 to 0.08 inches (1 to 2 millimeters). They eat a variety of foods that include eggs, carrion, insects, body fluids, nuts, seeds, grains, fruit nectar, sap and fungus.

Omnivores in the Desert

Jerboas, roadrunners and rabbit eared bandicoot are all omnivores in the desert ecosystem. The jerboa is a type of rodent that likes to eat plant seeds and insects. Roadrunners, a type of bird, sometimes likes to eat fruit but they also like to eat small rodents. The rabbit eared bandicoot also known as the bilby likes to eat fruits and plants but also likes to eat lizards and insects.

[Jerboa]

[Roadrunner]

[Rabbit Eared Bandicoot]

Omnivores in the Water

In the water there are many carnivores and herbivores but there are also some omnivores. Catfish, damselfish, and parrotfish are just some of the omnivores that live in the water.

[Catfish]

[Parrotfish]

[Damselfish]

These fish eat phytoplankton and other smaller fish like shrimp. There are also cleaner fish in the water, these fish feed on small plant and food pieces in the water but they also feed on parasites. Parasites are tiny animals that feed on larger animals for their nutrients, and many times they make the host animal sick.

Omnivores in the Forest

In the forest food chain different types of birds can be omnivores. Have you ever seen a bird in your front yard looking for a worm? Well that bird, whether it is a robin or a blue jay, is an omnivore because it eats seeds but it also eats worms.

[Bird]

[Mouse]

[Ants]

Other omnivores you see in the forest ecosystem are ants, and mice.

Omnivores in the Arctic

In the arctic, the grizzly bear is one type of omnivore. Scientists have found that 75% of grizzly bears can live off of plants alone. Though they seem to like plants these bears also like to eat small rodents, insects and honey. Arctic ground squirrels are also omnivores in the arctic, they like to eat most of any plant but they also like to eat insects, and carrion.

[Grizzly Bear]

[Ground Squirrel]

References

- Connie,, Rye,. Biology. Avissar, Yael,, Choi, Jung Ho, 1956-, DeSaix, Jean,, Jurukovski, Vladimir,, Wise, Robert R.,. Houston, Texas. ISBN 9781938168093. OCLC 896421272
- Peden, D. G.; Van Dyne, G. M.; Rice, R. W.; Hansen, R. M. (1974). "The Trophic Ecology of Bison bison L. on Shortgrass Plains". Journal of Applied Ecology. 11 (2): 489–497. doi:10.2307/2402203. JSTOR 2402203
- Dal Sasso, C.; Maganuco, S.; Cioffi, A. (26 May 2009). "A neurovascular cavity within the snout of the predatory dinosaur Spinosaurus". 1st International Congress on North African Vertebrate Palaeontology. Muséum national d'Histoire naturelle. Retrieved 22 September 2010
- Maron, John L.; Simms, Ellen L. (1997). "Effect of seed predation on seed bank size and seedling recruitment of bush lupine (Lupinus arboreus)". Oecologia. 111: 76. doi:10.1007/s004420050210
- Hulme, P.E. and Benkman, C.W. (2002) "Granivory", pp. 132–154 in Plant animal Interactions: An Evolutionary Approach, ed. C.M. Herrera and O. Pellmyr. Oxford: Blackwell. ISBN 978-0-632-05267-7
- Huber, D.R., Dean, M.N. and Summers, A.P., (2008), Hard prey, soft jaws and the ontogeny of feeding mechanics in the spotted ratfish Hydrolagus colliei. Journal of the Royal Society Interface(online publishing)
- Levermann, N., Galatius, A., Ehlme, G., Rysgaard, S. and Born, E.W. (2003). "Feeding behaviour of free-ranging walruses with notes on apparent dextrality of flipper use". BMC Ecology. 3 (9): 9. doi:10.1186/1472-6785-3-9. PMC 270045. PMID 14572316
- Karp, K. "Summary of Laboratory Experiments to Evaluate Consumption of Juvenile/Adult Quagga Mussel by Redear Sunfish and Bluegill". Retrieved June 8, 2013
- Ruppert, E.E., Fox, R.S. and Barnes, R.D., (2004). Invertebrate Zoology, 7th edition. Cengage Learning. ISBN 81-315-0104-3
- Buckley, Y. M., S. Anderson, C. P. Catterall, R. T. Corlett, T. Engel, C. R. Gosper, R. Nathan, D. M. Richardson, M. Setter, O. Spiegel, G. Vivian-Smith, F. A. Voigt, J. E. S. Weir, and D. A. Westcott. 2006. Management of plant invasions mediated by frugivore interactions. Journal of Applied Ecology 43:848-857
- Esselstyn, J.A., Achmadi, A.S. Rowe, K.C. (2012). Evolutionary novelty in a rat with no molars. Biology Letters, published online 22 August 2012, doi:10.1098/rsbl.2012.0574
- Steven L. Stephenson (21 April 2010). The Kingdom Fungi: The Biology of Mushrooms, Molds, and Lichens. Timber Press. pp. 200–. ISBN 9780881928914. Retrieved 10 February 2011

4

An Overview of Animal Intelligence

The study of mental cognition or intelligence involves a study of the different mental capacities of animals. Some of the animals whose intelligence is widely studied are dogs, cats, pigeons, elephants, etc. This chapter presents an overview of intelligence in such animals and also discusses significant topics, such as the theory of mind in animals and the mechanism of animal navigation.

The original method of studying animal cognition – also known as cognitive ethnology – was to perform experiments on intelligence in animals to determine the cognitive processes and patterns involved in animal behaviour.

This included things like classical conditioning and operant conditioning and was used to account for the more complex intellectual abilities exhibited by humans. However, this method only considered overt behaviour as valid data for study – in other words, only extreme forms of behaviour were considered as topics of interest.

The research methods were re-evaluated in the late 1900's and since then, researchers have begun to use a different approach: they start from known human mental processes and look for evidence comparable processes in other species. Most traditional research has always focused on rats and pigeons but researchers began to focus more on primates and cetaceans.

Their studies were also combined with results from studies of animal behaviour in the natural environment, via the fields of socio-biology, ethnology and behavioural ecology. These latter studies often show that animals need certain cognitive abilities in order to adapt to their ecological niche.

Areas of Animal Intelligence

In the search for animal analogs of human mental processes, researchers have concentrated their efforts in the following areas:

- Tools and Weapon Use – the use of tools for finding food and as weapons
- Categorisation – the ability to discriminate between different categories of stimuli

- Reasoning and Problem Solving – the ability to solve a range of problems, using abstract reasoning and supplying spontaneous solutions to problems without training
- Memory – the use of spatial memory for food catching and the possibility of short-term memory similar to that in humans.
- Consciousness – a hotly debated area, regarding whether animals have a concept of self and can display self-awareness.
- Attention - the ability to distribute attention between different aspects of a stimulus.
- Language – another controversial area, this involves the attempt to teach language or language-like behaviour to certain species
- Emotion – perhaps the most controversial of all, this involves research into whether animals have emotions or "feelings" which can be qualified in the same way as human feelings.

So do Animals Think Like us

Some of them may. It comes as no surprise that our closest biological relatives, the great apes, have been shown to be the most similar to us in intelligence, i.e.. their cognitive processes are similar to ours. However, bird species such as corvids and parrots, have also performed well against the standard.

Other animals reputed to have high human-like intelligence include the ceteceans (e.g. dolphins) but the difficulty and cost of research on these animals (as well as the small sample size) means that it is hard to provide conclusive supporting evidence. Similarly for cephalopods (e.g. octopuses) which are claimed to have higher-level problem-solving skills but again, there is only limited research in this area.

Relative Intelligence of different Animal Species

People have always viewed some animals as more intelligent than others: in European cultures, dogs, horses, great apes and (more recently) dolphins and parrots are seen as intelligent in ways that other animals are not. Crows have been attributed with humanlike intelligence by almost every culture that has encountered them. A number of recent survey studies have demonstrated the consistency of these rankings between people in a given culture. A common image is the *scala naturae*, the ladder of nature on which animals of different species occupy successively higher rungs, with humans typically at the top. Comparative psychologists have sought in vain for ways of providing an objective underpinning for these essentially subjective and anthropocentric judgements. Part of the difficulty is the lack of agreement about what we mean by intelligence even in humans (it obviously makes a big difference whether language is considered as essential for intelligence, for example). But in any case, different animals (including humans) seem to have different kinds of cognitive processes, which are better understood in terms of the ways in which they are cognitively adapted to their differentecological niches, than by positing any kind of hierarchy. One question that can be asked coherently is how far different species are intelligent in the same ways as humans are, i.e. are their cognitive processes similar to ours. Not surprisingly, our closest biological relatives, the great apes, tend to do best on such an assessment. It is less clear that the species traditionally held to be intelligent do unusually well against this standard, though among the birds, corvids and parrots typically are found to outperform other groups, and among the carnivores, dogs generally show better performance than cats. Despite ambitious claims, evidence of unusually high human-like intelligence among cetaceans is patchy, partly because the cost and difficulty of carrying out research with marine mammals mean that experiments frequently suffer from small sample sizes and inadequate controls and replication. Octopuses also exhibit many seemingly higher-level problem-solving skills.

Most Intelligent Animals on Earth

- Great apes (including Humans)
- New World monkeys
- Dolphin & Whales
- Dogs
- Octopi
- Birds, specifically Ravens and Crows
- Horses

Methods

The acceleration of research on animal cognition in the last 50 years or so has led to a rapid expansion in the variety of species studied and methods employed. The remarkable behavior of large-brained animals such as primates and cetacea has claimed special attention, but all sorts of mammals large and small, birds, fish, ants, bees, and others have been brought into the laboratory or observed in carefully controlled field studies. In the laboratory, animals push levers, pull strings, dig for food, swim in water mazes, or respond to images on computer screens in discrimination, attention, memory, and categorization experiments. Careful field studies explore memory for food caches, navigation by stars, communication, tool use, identification of conspecifics, and many other matters. Studies often focus on the behavior of animals in their natural environments and discuss the putative function of the behavior for the propagation and survival of the species. These developments reflect an increased cross-fertilization from related fields such as ethology and behavioral ecology. Also, contributions from behavioral neuroscience are beginning to clarify the physiological substrate of some inferred mental process.

Some researchers have made effective use of a Piagetian methodology, taking tasks which human children are known to master at different stages of development, and investigating which of them can be performed by particular species. Others have been inspired by concerns for animal welfare and the management of domestic species: for example Temple Grandin has harnessed her unique expertise in animal welfare and the ethical treatment of farm livestock to highlight underlying similarities between humans and other animals. From a methodological point of view, one of the main risks in this sort of work is anthropomorphism, the tendency to interpret an animal's behavior in terms of human feelings, thoughts, and motivations.

Research Questions

The common chimpanzee can use tools. This individual is using a stick to get food.

Human and animal cognition have much in common. Of course, research in the two also differs in important respects. Notably, much research with humans either studies or involves language, and much research with animals is related directly or indirectly to behaviors important to survival in natural settings.

Perception

Animals process information from eyes, ears, and other sensory organs to perceive the environment. Perceptual processes have been studied in many species, with results that are often similar to those in humans. Equally interesting are those perceptual processes that differ from, or go beyond those found in humans, such as echolocation in bats and dolphins, motion detection by skin receptors in fish, and extraordinary visual acuity, motion sensitivity and ability to see ultraviolet light in some birds.

Attention

Much of what is happening in the world at any moment is irrelevant to current behavior. Attention refers to mental processes that select relevant information, inhibit irrelevant information, and switch among these as the situation demands. Often the selective process is tuned before relevant information appears; such expectation makes for rapid selection of key stimuli when they become available. A large body of research has explored the way attention and expectation affect the behavior of non-human animals, and much of this work suggests that attention operates in birds, mammals and reptiles in much the same way that it does in humans.

Selective Learning

Animals trained to discriminate between two stimuli, say black versus white, can be said to attend to the "brightness dimension," but this says little about whether this dimension is selected in preference to others. More enlightenment comes from experiments that allow the animal to choose from several alternatives. For example, several studies have shown that performance is better on, for example, a color discrimination (e.g. blue vs green) after the animal has learned another color discrimination (e.g. red vs orange) than it is after training on a different dimension such as an X shape versus and O shape. The reverse effect happens after training on forms. Thus, the earlier learning appears to affect which dimension, color or form, the animal will attend to.

Other experiments have shown that after animals have learned to respond to one aspect of the environment responsiveness to other aspects is suppressed. In "blocking", for example, an animal is conditioned to respond to one stimulus ("A") by pairing that stimulus with reward or punishment. After the animal responds consistently to A, a second stimulus ("B") accompanies A on additional training trials. Later tests with the B stimulus alone elicit little response, suggesting that learning about B has been blocked by prior learning about A. This result supports the hypothesis that stimuli are neglected if they fail to provide new information. Thus, in the experiment just cited, the animal failed to attend to B because B added no information to that supplied by A. If true, this interpretation is an important insight into attentional processing, but this conclusion remains uncertain because blocking and several related phenomena can be explained by models of conditioning that do not invoke attention.

Divided Attention

Attention is a limited resource and is not an all-or-nothing response: the more attention devoted to one aspect of the environment, the less is available for others. A number of experiments have studied this in animals. In one experiment, a tone and a light are presented simultaneously to pigeons. The pigeons gain a reward only by choosing the correct combination of the two stimuli (e.g. a high frequency tone together with a yellow light). The birds perform well at this task, presumably by dividing attention between the two stimuli. When only one of the stimuli varies and the other is presented at its rewarded value, discrimination improves on the variable stimulus but discrimination on the alternative stimulus worsens. These outcomes are consistent with the notion that attention is a limited resource that can be more or less focused among incoming stimuli.

Visual Search and Attentional Priming

Attention functions to select information is of special use to the animal. Visual search typically calls for this sort of selection, and search tasks have been used extensively in both humans and animals to determine the characteristics of attentional selection and the factors that control it.

Experimental research on visual search in animals was initially prompted by field observations published by Luc Tinbergen (1960). Tinbergen observed that birds are selective when foraging for insects. For example, he found that birds tended to catch the same type of insect repeatedly even though several types were available. Tinbergen suggested that this prey selection was caused by an attentional bias that improved detection of one type of insect while suppressing detection of others. This "attentional priming" is commonly said to result from a pretrial activation of a mental representation of the attended object, which Tinbergen called a "searching image".

Tinbergen's field observations on priming have been supported by a number of experiments. For example, Pietrewicz and Kamil (1977, 1979) presented blue jays with pictures of tree trunks upon which rested either a moth of species A, a moth of species B, or no moth at all. The birds were rewarded for pecks at a picture showing a moth. Crucially, the probability with which a particular species of moth was detected was higher after repeated trials with that species (e.g. A, A, A,...) than it was after a mixture of trials (e.g. A, B, B, A, B, A, A...). These results suggest again that sequential encounters with an object can establish an attentional predisposition to see the object.

Another way to produce attentional priming in search is to provide an advance signal that is associated with the target. For example, if a person hears a song sparrow he or she may be predisposed to detect a song sparrow in a shrub, or among other birds. A number of experiments have reproduced this effect in animal subjects.

Still other experiments have explored nature of stimulus factors that affect the speed and accuracy of visual search. For example, the time taken to find a single target increases as the number of items in the visual field increases. This rise in RT is steep if the distracters are similar to the target, less steep if they are dissimilar, and may not occur if the distracters are very different in from the target in form or color.

Concepts and Categories

Fundamental but difficult to define, the concept of "concept" was discussed for hundreds of years by philosophers before it became a focus of psychological study. Concepts enable humans and animals to organize the world into functional groups; the groups may be composed of perceptually similar objects or events, diverse things that have a common function, relationships such as same versus different, or relations among relations such as analogies.

Methods

Most work on animal concepts has been done with visual stimuli, which can easily be constructed and presented in great variety, but auditory and other stimuli have been used as well. Pigeons have been widely used, for they have excellent vision and are readily conditioned to respond to visual targets; other birds and a number of other animals have been studied as well. In a typical experiment, a bird or other animal confronts a computer monitor on which a large number of pictures appear one by one, and the subject gets a reward for pecking or touching a picture of a category item and no reward for non-category items. Alternatively, a subject may be offered a choice between two or more pictures. Many experiments end with the presentation of items never seen before; successful sorting of these items shows that the animal has not simply learned many specific stimulus-response associations. A related method, sometimes used to study relational concepts, is matching-to-sample. In this task an animal sees one stimulus and then chooses between two or more alternatives, one of which is the same as the first; the animal is then rewarded for choosing the matching stimulus.

Perceptual Categories

Perceptual categorization is said to occur when a person or animal responds in a similar way to a range of stimuli that share common features. For example, a squirrel climbs a tree when it sees Rex, Shep, or Trixie, which suggests that it categorizes all three as something to avoid. This sorting of instances into groups is crucial to survival. Among other things, an animal must categorize if it is to apply learning about one object (e.g. Rex bit me) to new instances of that category (dogs may bite).

Natural Categories

Many animals readily classify objects by perceived differences in form or color. For example, bees or pigeons quickly learn to choose any red object and reject any green object if red leads to reward and green does not. Seemingly much more difficult is an animal's ability to categorize natural objects that vary a great deal in color and form even while belonging to the same group. In a classic study, Richard J. Herrnstein trained pigeons to respond to the presence or absence of human beings in photographs. The birds readily learned to peck photos that contained partial or full views of humans and to avoid pecking photos with no human, despite great differences in the form, size, and color of both the humans displayed and in the non-human pictures. In follow-up studies, pigeons categorized other natural objects (e.g. trees) and after training they were able without reward to sort photos they had not seen before. Similar work has been done with natural auditory categories, for example, bird songs. Honeybees (*Apis mellifera*) are able to form concepts of "up" and "down".

Functional or Associative Categories

Perceptually unrelated stimuli may come to be responded to as members of a class if they have a common use or lead to common consequences. An oft cited study by Vaughan (1988) provides an example. Vaughan divided a large set of unrelated pictures into two arbitrary sets, A and B. Pigeons got food for pecking at pictures in set A but not for pecks at pictures in set B. After they had learned this task fairly well, the outcome was reversed: items in set B led to food and items in set A did not. Then the outcome was reversed again, and then again, and so on. Vaughan found that after 20 or more reversals, associating reward with a few pictures in one set caused the birds to respond to the other pictures in that set without further reward, as if they were thinking "if these pictures in set A bring food, the others in set A must also bring food." That is, the birds now categorized the pictures in each set as functionally equivalent. Several other procedures have yielded similar results.

Relational or Abstract Categories

When tested in a simple stimulus matching-to-sample task many animals readily learn specific item combinations, such as "touch red if the sample is red, touch green if the sample is green." But this does not demonstrate that they distinguish between "same" and "different" as general concepts. Better evidence is provided if, after training, an animal successfully makes a choice that matches a novel sample that it has never seen before. Monkeys and chimpanzees do learn to do this, as do pigeons if they are given a great deal of practice with many different stimuli. However, because the sample is presented first, successful matching might mean that the animal is simply choosing the most recently seen "familiar" item rather than the conceptually "same" item. A number of studies have attempted to distinguish these possibilities, with mixed results.

Rule Learning

The use of rules has sometimes been considered an ability restricted to humans, but a number of experiments have shown evidence of simple rule learning in primates and also in other animals. Much of the evidence has come from studies of sequence learning in which the "rule" consists of the order in which a series of events occurs. Rule use is shown if the animal learns to discriminate different orders of events and transfers this discrimination to new events arranged in the same order. For example, Murphy *et al.* (2008) trained rats to discriminate between visual sequences. For one group ABA and BAB were rewarded, where A="bright light" and B="dim light". Other stimulus triplets were not rewarded. The rats learned the visual sequence, although both bright and dim lights were equally associated with reward. More importantly, in a second experiment with auditory stimuli, rats responded correctly to sequences of novel stimuli that were arranged in the same order as those previously learned. Similar sequence learning has been demonstrated in birds and other animals as well.

Memory

The categories that have been developed to analyze human memory (short term memory, long term memory, working memory) have been applied to the study of animal memory, and some of the phenomena characteristic of human short term memory (e.g. the serial position effect) have been detected in animals, particularly monkeys. However most progress has been made in the analysis of spatial memory; some of this work has sought to clarify the physiological basis of spatial memory and the role of the hippocampus; other work has explored the spatial memory of scatter-hoarder animals such as Clark's nutcracker, certain jays, tits and certain squirrels, whose ecological niches require them to remember the locations of thousands of caches, often following radical changes in the environment.

Memory has been widely investigated in foraging honeybees, *Apis mellifera*, which use both transient short-term working memory that is non-feeder specific and a feeder specific long-term reference memory. Memory induced in a free-flying honeybee by a single learning trial lasts for days and, by three learning trials, for a lifetime. Slugs, *Limax flavus*, have a short-term memory of approximately 1 min and long-term memory of 1 month.

Methods

As in humans, research with animals distinguishes between "working" or "short-term" memory from "reference" or long-term memory. Tests of working memory evaluate memory for events that happened in the recent past, usually within the last few seconds or minutes. Tests of reference memory evaluate memory for regularities such as "pressing a lever brings food" or "children give me peanuts".

Habituation

This is one of the simplest tests for memory spanning a short time interval. The test compares an animal's response to a stimulus or event on one occasion to its response on a previous occasion. If the second response differs consistently from the first, the animal must have remembered something about the first, unless some other factor such as motivation, sensory sensitivity, or the test stimulus has changed.

Delayed Response

Delayed response tasks are often used to study short-term memory in animals. Introduced by Hunter (1913), a typical delayed response task presents an animal with a stimulus such a colored light, and after a short time interval the animal chooses among alternatives that match the stimulus, or are related to the stimulus in some other way. In Hunter's studies, for example, a light appeared briefly in one of three goal boxes and then later the animal chose among the boxes, finding food behind the one that had been lighted. Most research has been done with some variation of the "delayed matching-to-sample" task. For example, in the initial study with this task, a pigeon was presented with a flickering or steady light. Then, a few seconds later, two pecking keys were illuminated, one with a steady light and one with a flickering light. The bird got food if it pecked the key that matched the original stimulus.

A commonly-used variation of the matching-to-sample task requires the animal to use the initial stimulus to control a later choice between different stimuli. For example, if the initial stimulus is a black circle, the animal learns to choose "red" after the delay; if it is a black square, the correct choice is "green". Ingenious variations of this method have been used to explore many aspects of memory, including forgetting due to interference and memory for multiple items.

Radial Arm Maze

The radial arm maze is used to test memory for spatial location and to determine the mental processes by which location is determined. In a radial maze test, an animal is placed on a small platform from which paths lead in various directions to goal boxes; the animal finds food in one or more goal boxes. Having found food in a box, the animal must return to the central platform. The maze may be used to test both reference and working memory. Suppose, for example, that over a number of sessions the same 4 arms of an 8-arm maze always lead to food. If in a later test session the animal goes to a box that has never been baited, this indicates a failure of reference memory. On the other hand, if the animal goes to a box that it has already emptied during the same test session, this indicates a failure of working memory. Various confounding factors, such as odor cues, are carefully controlled in such experiments.

Water Maze

The water maze is used to test an animal's memory for spatial location and to discover how an animal is able to determine locations. Typically the maze is a circular tank filled with water that has been made milky so that it is opaque. Located somewhere in the maze is small platform placed just below the surface of the water. When placed in the tank, the animal swims around until it finds and climbs up on the platform. With practice, the animal finds the platform more and more quickly. Reference memory is assessed by removing the platform and observing the relative amount of time the animal spends swimming in the area where the platform had been located. Visual and other cues in and around the tank may be varied to assess the animal's reliance on landmarks and the geometric relations among them.

Spatial Cognition

Whether an animal ranges over a territory of measured in square kilometers or square meters, its survival typically depends on its ability to do such things as find a food source and then return to its nest. Sometimes such a task can be performed rather simply, for example by following a chemical trail. Typically, however, the animal must somehow acquire and use information about locations, directions, and distances. The following paragraphs outline some of the ways that animals do this.

- Beacons: Animals often learn what their nest or other goal looks like, and if it is within sight they may simply move toward it; it is said to serve as a "beacon".
- Landmarks: When an animal is unable to see its goal, it may learn the appearance of nearby objects and use these landmarks as guides. Researchers working with birds and bees have demonstrated this by moving prominent objects in the vicinity of nest sites, causing returning foragers to hunt for their nest in a new location.
- Dead reckoning: Dead reckoning, also known as "path integration," is the process of computing one's position by starting from a known location and keeping track of the distances and directions subsequently traveled. Classic experiments have shown that the desert ant keeps track of its position in this way as it wanders for many meters searching for food. Though it travels in a randomly twisted path, it heads straight home when it finds food. However, if the ant is picked up and released some meters to the east, for example, it heads for a location displaced by the same amount to the east of its home nest.
- Cognitive maps: Some animals appear to construct a cognitive map of their surroundings, meaning that they acquire and use information that enables them to compute how far and in what direction to go to get from one location to another. Such a map-like representation is thought to be used, for example, when an animal goes directly from one food source to another even though its previous experience has involved only travel between each source and home. Research in this area has also explored such topics as the use of geometric properties of the environment by rats and pigeons, and the ability of rats to represent a spatial pattern in either radial arm mazes or water mazes. Spatial cognition is sometimes explored in visual search experiments in which a human or animal searches the environment for a particular object.

- Detour behaviour: Some animals appear to have an advanced understanding of their spatial environment and will not take the most direct route if this confers an advantage to them. Some jumping spiders take an indirect route to prey rather than the most direct route, thereby indicating flexibility in behaviour and route planning, and possibly insight learning.

Long-distance Navigation; Homing

Many animals travel hundreds or thousands of miles in seasonal migrations or returns to breeding grounds. They may be guided by the sun, the stars, the polarization of light, magnetic cues, olfactory cues, winds, or a combination of these.

It has been hypothesized that animals such as apes and wolves are good at spatial cognition because this skill is necessary for survival. Some researchers argue that this ability may have diminished somewhat in dogs because humans have provided necessities such as food and shelter during some 15,000 years of domestication.

Timing

Time of Day: Circadian Rhythms

The behavior of most animals is synchronized with the earth's daily light-dark cycle. Thus, many animals are active during the day, others are active at night, still others near dawn and dusk. Though one might think that these "circadian rhythms" are controlled simply by the presence or absence of light, nearly every animal that has been studied has been shown to have a "biological clock" that yields cycles of activity even when the animal is in constant illumination or darkness. Circadian rhythms are so automatic and fundamental to living things – they occur even in plants – that they are usually discussed separately from cognitive processes.

Interval Timing

Survival often depends on an animal's ability to time intervals. For example, rufous hummingbirds feed on the nectar of flowers, and they often return to the same flower, but only after the flower has had enough time to replenish its supply of nectar. In one experiment hummingbirds fed on artificial flowers that quickly emptied of nectar but were refilled at some fixed time (e.g. twenty minutes) later. The birds learned to come back to the flowers at about the right time, learning the refill rates of up to eight separate flowers and remembering how long ago they had visited each one.

The details of interval timing have been studied in a number of species. One of the most common methods is the "peak procedure". In a typical experiment, a rat in an operant chamber presses a lever for food. A light comes on, a lever-press brings a food pellet at a fixed later time, say 10 seconds, and then the light goes off. Timing is measured during occasional test trials on which no food is presented and the light stays on. On these test trials, the rat presses the lever more and more until about 10 sec and then, when no food comes, gradually stops pressing. The time at which the rat presses most on these test trials is taken to be its estimate of the payoff time.

Experiments using the peak procedure and other methods have shown that animals can time short intervals quite exactly, can time more than one event at once, and can integrate time with spatial and other cues. Such tests have also been used for quantitative tests of theories of animal timing, such as Gibbon's Scalar Expectancy Theory ("SET"), Killeen's Behavioral Theory of Timing, and Machado's Learning to Time model. No one theory has yet gained unanimous agreement.

Tool and Weapon Use

Although tool use was long assumed to be a uniquely human trait, there is now much evidence that many animals use tools, including mammals, birds, fish, cephalopods and insects. Discussions of tool use often involve a debate about what constitutes a "tool", and they often consider the relation of tool use to the animal's intelligence and brain size.

Mammals

Series of photographs showing a bonobo fishing for termites.

A bonobo inserting a stick into a termite mound.

The bonobo starts "fishing" for the termites.

The bonobo withdraws the stick and begins eating the termites.

The bonobo eats the termites extracted with the tool.

Tool use has been reported many times in both wild and captive primates, particularly the great apes. The use of tools by primates is varied and includes hunting (mammals, invertebrates, fish), collecting honey, processing food (nuts, fruits, vegetables and seeds), collecting water, weapons and shelter. Research in 2007 shows that chimpanzees in the Fongoli savannah sharpen sticks to use as spears when hunting, considered the first evidence of systematic use of weapons in a species other than humans. Other mammals that spontaneously use tools in the wild or in captivity include elephants, bears, cetaceans, sea otters and mongooses.

Birds

Several species of birds have been observed to use tools in the wild, including warblers, parrots, Egyptian vultures, brown-headed nuthatches, gulls and owls. Some species, such as the woodpecker finch of the Galapagos Islands, use particular tools as an essential part of their foraging behavior. However, these behaviors are often quite inflexible and cannot be applied effectively in new situations. A great many species of birds build nests with a wide range of complexities, but although nest-building behaviour fulfills the criteria of some definitions of "tool-use", this is not the case with other definitions.

Several species of corvids have been trained to use tools in controlled experiments. One species examined extensively under laboratory conditions is the New Caledonian crow. One individual called "Betty" spontaneously made a wire tool to solve a novel problem. She was being tested to see whether she would select a wire hook rather than a straight wire to pull a little bucket of meat out of a well. Betty tried poking the straight wire at the meat. After a series of failures with this direct approach, she withdrew the wire and began directing it at the bottom of the well, which was secured to its base with duct tape. The wire soon became stuck, whereupon Betty pulled it sideways, bending it and unsticking it. She then inserted the hook into the well and extracted the meat. In all but one of 10 subsequent trials with only straight wire provided, she also made and used a hook in the same manner, but not before trying the straight wire first.

Fish

Several species of wrasses have been observed using rocks as anvils to crack bivalve (scallops, urchins and clams) shells. This behavior was first filmed in an orange-dotted tuskfish (*Choerodon anchorago*) in 2009 by Giacomo Bernardi. The fish fans sand to unearth the bivalve, takes it into its mouth, swims several meters to a rock, which it then uses as an anvil by smashing the mollusc apart with sideward thrashes of the head. This behaviour has also been recorded in a blackspot tuskfish (*Choerodon schoenleinii*) on Australia's Great Barrier Reef, yellowhead wrasse (*Halichoeres garnoti*) in Florida and a six-bar wrasse (*Thalassoma hardwicke*) in an aquarium setting. These species are at opposite ends of the phylogenetic tree in this family, so this behaviour may be a deep-seated trait in all wrasses.

Invertebrates

Some cephalopods are known to use coconut shells for protection or camouflage.

Ants of the species *Conomyrma bicolor* pick up stones and other small objects with their mandibles and drop them down the vertical entrances of rival colonies, allowing workers to forage for food without competition.

Reasoning and Problem Solving

It is clear that animals of quite a range of species are capable of solving problems that appear to require abstract reasoning; Wolfgang Köhler's (1917) work with chimpanzees is a famous early example. He observed that chimpanzees did not use trial and error to solve problems such as retrieving bananas hung out of reach. Instead, they behaved in a manner that was "unwaveringly purposeful," spontaneously placing boxes so that they could climb to reach the fruit. Modern research has identified similar behavior in animals usually thought of as much less intelligent, if appropriate pre-training is given. Causal reasoning has also been observed in rooks and New Caledonian crows.

It has been shown that Barbados bullfinches (*Loxigilla barbadensis*) from urbanized areas are better at innovative problem-solving tasks than bullfinches from rural environments, but that they did not differ in colour discrimination learning.

Cognitive Bias

Is the glass half empty or half full?

A cognitive bias refers to a systematic pattern of deviation from norm or rationality in judgment, whereby inferences about other individuals or situations may be drawn in an illogical fashion.

Cognitive bias is sometimes illustrated by using answers to the question "Is the glass half empty or half full?". Choosing "half empty" is supposed to indicate pessimism whereas choosing "half full" indicates optimism. To test this in animals, an individual is trained to anticipate that stimulus A, e.g. a 100 Hz tone, precedes a positive event, e.g. highly desired food is delivered when a lever is pressed by the animal. The same individual is trained to anticipate that stimulus B, e.g. a 900 Hz tone, precedes a negative event, e.g. bland food is delivered when the animal presses a lever. The animal is then tested by being given an intermediate stimulus C, e.g. a 500 Hz tone, and observing whether the animal presses the lever associated with the positive or negative reward. This has been suggested to indicate whether the animal is in a positive or negative mood.

In a study that used this approach, rats that were playfully tickled responded differently than rats that were simply handled. The rats that had been tickled were more optimistic than the handled rats. The authors suggested that they had demonstrated "...for the first time a link between the directly measured positive affective state and decision making under uncertainty in an animal model".

There is some evidence for cognitive bias in a number of species, including rats, dogs, rhesus macaques, sheep, chicks, starlings and honeybees.

Language

The modeling of human language in animals is known as animal language research. In addition to the ape-language experiments mentioned above, there have also been more or less successful attempts to teach language or language-like behavior to some non-primate species, including parrots and great spotted woodpeckers. Arguing from his own results with the animal Nim Chimpsky and his analysis of others results, Herbert Terrace criticized the idea that chimps can produce new sentences. Shortly thereafter Louis Herman published research on artificial language comprehension in the bottlenosed dolphin (Herman, Richards, & Wolz, 1984). Though this sort of research has been controversial, especially among cognitive linguists, many researchers agree that many animals can understand the meaning of individual words, and that some may understand simple sentences and syntactic variations, but there is little evidence that any animal can produce new strings of symbols that correspond to new sentences.

Insight

Wolfgang Köhler is usually credited with introducing the concept of insight into experimental psychology. Working with chimpanzees, Köhler came to dispute Edward Thorndike's theory that animals must solve problems gradually, by trial and error. He said that Thorndike's animals could only use trial and error because the situation precluded other forms of problem solving. He provided chimps with a relatively unstructured situation, and he observed sudden "ah-ha!" insightful changes of behavior, as, for example, when a chimp suddenly moved a box into position so that it could retrieve a banana. More recently, Asian elephants (*Elephas maximus*) were shown to exhibit similar insightful problem solving. A male was observed moving a box to a position where it could be stood upon to reach food that had been deliberately hung out of reach.

Numeracy

A variety of studies indicates that animals are able to use and communicate quantitative information, and that some can count in a rudimentary way. Some examples of this research follow.

In one study, rhesus monkeys viewed visual displays containing, for example, 1,2, 3, or 4 items of different sorts. They were trained to respond to them in several ways involving numerical ordering, for example touching "1" first, "2" second and so on. When tested with displays containing items they had never seen before, they continued to respond to them in order. The authors conclude that monkeys can represent the numerosities 1 to 9 at least on an ordinal scale.

Ants are able to use quantitative values and transmit this information. For instance, ants of several species are able to estimate quite precisely numbers of encounters with members of other colonies on their feeding territories. Numeracy has been described in the yellow mealworm beetle (*Tenebrio molitor*) and the honeybee.

Western lowland gorillas given the choice between two food trays demonstrated the ability to choose the tray with more food items at a rate higher than chance after training. In a similar task, chimpanzees chose the option with the larger amount of food. Salamanders given a choice between two displays with differing amounts of fruit flies, used as a food reward, reliably choose the display with more flies, as shown in a particular experiment.

Other experiments have been conducted that show animals' abilities to differentiate between non-food quantities. American black bears demonstrated quantity differentiation abilities in a task with a computer screen. The bears were trained to touch a computer monitor with a paw or nose to choose a quantity of dots in one of two boxes on the screen. Each bear was trained with reinforcement to pick a larger or smaller amount. During training, the bears were rewarded with food for a correct response. All bears performed better than what random error predicted on the trials with static, non-moving dots, indicating that they could differentiate between the two quantities. The bears choosing correctly in congruent (number of dots coincided with area of the dots) and incongruent (number of dots did not coincide with area of the dots) trials suggests that they were indeed choosing between quantities that appeared on the screen, not just a larger or smaller retinal image, which would indicate they are only judging size.

Bottlenose dolphins have shown the ability to choose an array with fewer dots compared to one with more dots. Experimenters set up two boards showing various numbers of dots in a poolside setup. The dolphins were initially trained to choose the board with the fewer number of dots. This was done by rewarding the dolphin when it chose the board with the fewer number of dots. In the experimental trials, two boards were set up, and the dolphin would emerge from the water and point to one board. The dolphins chose the arrays with fewer dots at a rate much larger than chance, indicating they can differentiate between quantities. A particular grey parrot, after training, has shown the ability to differentiate between the numbers zero through six using vocalizations. After number and vocalization training, this was done by asking the parrot how many objects there were in a display. The parrot was able to identify the correct amount at a rate higher than chance. Angelfish, when put in an unfamiliar environment will group together with conspecifics, an action named shoaling. Given the choice between two groups of differing size, the angelfish will choose the larger of the two groups. This can be seen with a discrimination ratio of 2:1 or greater, such that, as long as one group has at least twice the fish as another group, it will join the larger one.

Monitor lizards have been shown to be capable of numeracy, and some species can distinguish among numbers up to six.

Intelligence

As the Cognitive ability and intelligence in non-human animals cannot be measured with verbal scales, it has been measured using a variety of methods that involve such things as habit reversal, social learning, and responses to novelty. Principal Component Analysis and factor analytic studies have shown that a single factor of intelligence is responsible for 47% of the individual variance in cognitive ability measures in primates and between 55% and 60% of the variance in mice. These values are similar to the accepted variance in IQ explained by a similar single factor known as the general factor of intelligence in humans (40-50%).

The general factor of intelligence, or *g* factor, is a psychometric construct that summarizes the correlations observed between an individual's scores on various measures of cognitive abilities. It has been suggested that *g* is related to evolutionary life histories and the evolution of intelligence as well as to social learning and cultural intelligence. Non-human models of *g* have been used in genetic and neurological research on intelligence to help understand the mechanisms behind variation in *g*.

Theory of Mind

Theory of mind is the ability to attribute mental states, e.g. intents, desires, pretending, knowledge, to oneself and others and to understand that others have desires, intentions, and perspectives that are different from one's own.

Some research with ravens provides an example of evidence for theory of mind in a non-human species. Ravens are members of the corvidae family, which is widely regarded as having high cognitive abilities. These birds have been observed to hide their food when dominant ravens are visible and audible at the same time. Based on this observation, ravens were tested for their understanding of "seeing" as a mental state. In a first step, the birds protected their cache when dominants were visible but not when they could only be heard from an adjacent room. In the next step, they had access to a small peephole which allowed them to see into the adjacent room. With the peephole open, the ravens guarded their caches against discovery when they could hear dominants in the adjacent room, even when the dominant's sounds were playbacks of recordings.

Consciousness

Mirror test with a baboon

The sense in which animals can be said to have consciousness or a self-concept has been hotly debated. The best known research technique in this area is the mirror test devised by Gordon G. Gallup, in which an animal's skin is marked in some way while it is asleep or sedated, and it is then allowed to see its reflection in a mirror; if the animal spontaneously directs grooming behavior towards the mark, that is taken as an indication that it is aware of itself. Self-awareness, by this criterion, has been reported for chimpanzees and also for other great apes, the European magpie, some cetaceans and an Asian elephant , but not for monkeys. The mirror test has been criticized by researchers because it is entirely focused on vision, the primary sense in humans, while other species rely more heavily on other senses such as the olfactory sense in dogs.

It has been suggested that metacognition in some animals provides some evidence for cognitive self-awareness. The great apes, dolphins, and rhesus monkeys have demonstrated the ability to monitor their own mental states and use an "I don't know" response to avoid answering difficult questions. Unlike the mirror test, which reveals awareness of the condition of one's own body, this uncertainty monitoring is thought to reveal awareness of one's internal mental state. A 2007 study has provided some evidence for metacognition in rats, although this interpretation has been questioned. These species might also be aware of the strength of their memories.

Some researchers propose that animal calls and other vocal behaviors provide evidence of consciousness. This idea arose from research on children's crib talk by Weir (1962) and in investigations of early speech in children by Greenfield and others (1976). Some such research has been done with a macaw.

In July, 2012 during the "Consciousness in Human and Nonhuman Animals" conference in Cambridge a group of scientists announced and signed a declaration with the following conclusions:

> Convergent evidence indicates that non-human animals have the neuroanatomical, neurochemical, and neurophysiological substrates of conscious states along with the capacity to exhibit intentional behaviors. Consequently, the weight of evidence indicates that humans are not unique in possessing the neurological substrates that generate consciousness. Non-human animals, including all mammals and birds, and many other creatures, including octopuses, also possess these neurological substrates.

Biological Constraints

Hedgehogs instinctively roll into a ball when threatened, making them unsuitable for studies on aversion avoidance

Animals differ widely in many learning and cognitive tasks in ways that reflect their evolutionary history and their instinctual behaviors in natural environments. For example, dogs and rats easily learn to avoid an electric shock from the floor by moving to another part of the experimental chamber when they hear a tone preceding the shock; this is an appropriate response to a dangerous situation. However, hedgehogs fail to learn this avoidance behavior. This might seem to show the hedgehog's inability to learn, but the hedgehog's instinctive reaction to a threat is to curl up into a ball, a response that interferes with possible escape behavior in this situation.

Instinctive drift is another factor that can influence the interpretation of cognitive research. Instinctive drift is the tendency of an animal to revert to instinctive behaviors that can interfere with learned responses. The concept originated with Keller and Marian Breland when they taught a raccoon to put coins into a box. The raccoon drifted to its instinctive behavior of rubbing the coins with its paws, as it would do when foraging for food.

Animal ability to process and respond to stimuli is correlated with brain size. Small-brain animals tend to show simple behaviors that are less dependent on learning than those of large-brained animals. Vertebrates, particularly mammals, have large brains and complex behavior that changes with experience. A formula called the encephalization quotient (EC) expresses a relationship between brain and body size; it was developed by H.J. Jerison in the late 1960s. When the encephalization quotient is plotted as a curve, an animal with an EC above the curve is expected to show more cognitive ability than the average animal of its size, whereas an animal with an EC below the curve is expected to have less. Various formulas been suggested, but the equation $Ew(brain) = 0.12w(body)^{2/3}$ has been found to fit data from a sample of mammals. The formula is suggestive at best, and should only be applied to non-mammals with extreme caution. For some of the other vertebrate classes, the power of 3/4 rather than 2/3 is sometimes used, and for many groups of invertebrates, the formula may not give meaningful results.

Cognitive Faculty by Species

A traditionally common image is the *scala naturae*, the ladder of nature on which animals of different species occupy successively higher rungs, with humans typically at the top. However, rather than using such an arbitrary hierarchy, it seems more fruitful to understand cognitive capacities as adaptations to differing ecological niches.

Whether fairly or not, the performance of animals is often compared to that of humans on cognitive tasks. Not surprisingly, our closest biological relatives, the great apes, tend to perform most like humans. Among the birds, corvids and parrots have typically been found to perform well on human-like tasks. Some octopodes have also been shown to exhibit a number of higher-level skills such as tool use, but the amount of research on cephalopod intelligence is still limited

Baboons have been shown to be capable of recognizing words.

Elephant Intelligence

Elephants continue to fascinate both scientists and general observers alike. They are recognised as being among the most intelligent creatures on earth. In fact, some enthusiasts believe that their intelligence rivals that of human beings.

Aristotle even said of elephants: "The beast which passeth all others in wit and mind".

Proportionally, the elephant's brain is the most sizeable at a mass of just over 5kg. Although the largest whale is 20 times the body size of an elephant, its brain is just under twice the size. The need for such a large and complex organ becomes clear when we consider the behaviours and abilities of these animals. Elephants are capable of a range of emotions, including joy, playfulness, grief and mourning. In addition, elephants are able to learn new facts and behaviours, mimic sounds that they hear, self-medicate, play with a sense of humour, perform artistic activities, use tools and display compassion and self-awareness.

Part of the reason that elephants possess such a superior level of intelligence is the structure of their brain. Their neocortex is highly convoluted, as it is in humans, apes and some dolphins. This is generally accepted to be an indication of complex intelligence. The cortex is thick and comprises many neurons. The elephant is one of the few creatures (along with human beings) that is not born with survival instincts, but needs to learn these during infancy and adolescence. The brain is specially designed to accomplish this sort of life learning. Elephants and humans have a similar lifespan, and plenty of time, approximately 10 years, is allowed for them to learn before they are considered to be independent adults.

The lessons learnt include how to feed, use tools and understand their place in their social structure. Elephants capacity for memory and emotions is remarkable and is due to the well-developed hippocampus. This is also the area responsible for emotional flashbacks and is the reason that elephants experience Post-Traumatic Stress Disorder.

The insight and intelligence of the elephant is particularly note-worthy in their ability to mourn their dead. This behaviour has only previously been noted in humans. In fact, recently deceased elephants will receive a burial ceremony, while those who are already reduced to a skeleton are still paid respect by passing herds. The burial ceremony is marked by deep rumblings while the dead body is touched and caressed by the herd members trunks.

Intelligence is also manifested in the elephant's ability to self-medicate. When a pregnant mother is due to give birth, she will chew on the leaves of the tree from the Boraginaceae family to induce labour.

Another ability that indicates superior intellect is elephants ability to play and display a sense of humour. Games include throwing a stick at a certain object, passing an object from one animal to another, or squirting water out of the trunk in a fountain. Elephants in zoos have even been seen stealing onlookers caps and hiding them in playful teasing.

The ability to mimic sounds is another indication of the impressive intelligence of these beasts. Elephants have been recorded mimicking passing trucks and even the sounds made by their trainers. Often, the elephant manages to articulate certain sounds so that they bear a strong resemblance to the spoken word.

Elephants are able to use tools or implements to accomplish a task they cannot perform on their own. They have been observed digging holes for drinking water, then moulding bark from a tree into the shape of a ball and placing it on top of the hole and covering it over with sand to avoid evaporation. They also use sticks to scratch their backs when their trunk can not reach and have been known to drop rocks on electric fences to damage them.

The elephant's problem solving abilities are another impressive facet of their boundless intelligence. Incredibly, the elephant is able to change its behaviour based on a given situation. Bandula, an Asian elephant in captivity, had learnt how to release the complex hook on her shackles and would then assist her fellow "inmates" to escape from theirs.

Self awareness is yet another indication of the vast capacity for thinking and intellect that exists in the elephant. They can, in fact, recognise themselves in a mirror, something that is extremely rare in the animal kingdom.

These capabilities are merely touching the tip of the iceberg of what is the elephant's capacity for insight, thought and discernment. And it is this capacity that continues to captivate researchers and onlookers alike in their eternal quest to understand the mystery of the elephant psyche.

Structure of the Brain

Cerebral Cortex

The elephant (both Asian and African) has a very large and highly complex neocortex, a trait also shared by humans, other apes and certain dolphin species.

Asian elephants have the greatest volume of cerebral cortex available for cognitive processing of all existing land animals. Elephants have a volume of cerebral cortex available for cognitive processing that exceeds that of any primate species, with one study suggesting elephants be placed in the category of great apes in terms of cognitive abilities for tool use and tool making.

The elephant brain exhibits a gyral pattern more complex and with more numerous convolutions, or brain folds, than that of humans, other primates, or carnivores, but less complex than cetaceans, although elephants have as many cortical neurons (nerve cells) and cortical synapses as that of humans, which is more than that of the cetaceans such as whales and dolphins. Elephants are believed to rank equal with dolphins in terms of problem-solving abilities, and many scientists tend to rank elephant intelligence at the same level as cetaceans; a 2011 article published by ABC Science states that, "elephants [are as] smart as chimps, [and] dolphins".

Other Areas of the Brain

Elephants also have a very large and highly convoluted hippocampus, a brain structure in the limbic system that is much bigger than that of any human, primate or cetacean. The hippocampus of an elephant takes up about 0.7% of the central structures of the brain, comparable to 0.5% for humans and with 0.1% in Risso's dolphins and 0.05% in bottlenose dolphins.

The hippocampus is linked to emotion through the processing of certain types of memory, especially spatial. This is thought to be possibly why elephants suffer from psychological flashbacks and the equivalent of post-traumatic stress disorder (PTSD).

The encephalization quotient (EQ) (the size of the brain relative to body size) of elephants ranges from 1.13 to 2.36. The average EQ is 2.14 for Asian elephants, and 1.67 for African, with the overall average being 1.88. In comparison to other animals, the La Plata dolphin has an EQ of 1.67; the Ganges River dolphin of 1.55; the orca of 2.57; the bottlenose dolphin of 4.14; and the tucuxi dolphin of 4.56; chimpanzees at 2.49; dogs at 1.17, cats at 1.00; and mice at 0.50. Humans have an EQ of 7.44.

Brain Size at Birth Relative to Adult Brain Size

Like humans, rather than instinctual, elephants must learn most survival behavior and strategy as they grow up. Elephants have a very long period in their lives for learning, lasting for around ten years. One comparative way to try to gauge intelligence is to compare brain size at birth to the fully developed adult brain. This indicates how much learning a species accumulates while young. The majority of mammals are born with a brain close to 90% of the adult weight, while humans are born with 28%, bottlenose dolphins with 42.5%, chimpanzees with 54%, and elephants with 35%. This indicates that elephants have the highest amount of learning to undergo next to humans, and behavior is not mere instinct but must be taught throughout life. It should be noted that instinct is quite different from learned intelligence. Parents teach their young how to feed, use tools and learn their place in the highly complex elephant society. The cerebrum temporal lobes, which function as storage of memory, are much larger than those of a human.

Spindle Neurons

Spindle cells appear to play a central role in the development of intelligent behavior. As well as in humans and the rest of the great apes, spindle neurons are also found in the brains of both Asian and African elephants, as well as humpback whales, fin whales, killer whales, sperm whales, bottlenose dolphins, Risso's dolphins, and beluga whales. The remarkable similarity between the elephant brain and the human brain supports the thesis of convergent evolution.

Discrimination Tasks

Thorndike argued that his cats and dogs escaped puzzle boxes through a mindless process of trial and error. Because understanding something as simple as pulling a loop to open a door must occur rapidly or not at all, it should have induced, at some point during the repeated introductions of his animals into the box, a sudden reduction in escape time. The actual, gradual, slope of the time-curve that he did observe suggested to him that his subjects failed to understand the cause-effect relationships between their actions and escape.

In 1957, Altevogt and Rensch reported that a young Asian elephant needed 330 trials, over a period of several days, to consistently choose the reinforced response in her first discrimination task. In an experiment which employed another sense modality, an 8-year-old took 7.5 months to distinguish 12 tones.

Similarly, in discrimination experiments with Asian elephants in the logging camps of Myanmar, only 13 Burmese elephants mastered black/white or large/small visual discrimination tasks, while 7 elephants failed to master the task.

Like Thorndike's cats and dogs, the 13 elephants that did master a black/white or large/small discrimination task did so gradually, over several sessions. The slope of the time-curve again suggested the elephants failed to understand the cause-effect relationships between lid removal and food retrieval.

The pre-training that preceded these Myanmar discrimination experiments involved learning to remove a lid from a bucket or to displace a box to uncover a hole in the ground. On average, "the 20 elephants taking part in these experiments required 3.4 sessions to gradually master the pre-training task."

Possible Absence of Causal Reasoning

Thorndike's dogs and cats learned to escape a box by, for instance, pulling a loop attached to a string that opened the box. On subsequent introductions to an open box when string-pulling no longer served a purpose, the animal continued to pull a string before getting out of the box. Thorndike concluded that the animal solved the task mechanically, without understanding the causal link between string-pulling and escaping.

A conceptually similar experiment involved pre-training four logging Asian elephants to remove food from a cover-less bucket by inserting their trunk into the bucket. Next, a treat was placed at the bottom of a bucket and, at the same time, the bucket was covered with a lid. The elephants were then trained to obtain the treat by removing the lid. Once this behavioral sequence was established, a treat was placed inside the bucket and, at the same time, the lid was placed on the ground alongside the bucket, so that the lid no longer obstructed access to the food.

As in the Thorndike case, if the elephants understand the nature of the task, they should ignore the lid on the ground and retrieve the reward directly, as they used to do in pre-training sessions before the lid was introduced. But if they fail to understand the causal link between lid removal and food retrieval, they might continue to remove the lid before retrieving the food. The observations accorded entirely with Thorndike's mechanical learning hypothesis.

Dog Intelligence

Dog intelligence is the ability of a dog to learn, think, and solve problems. Dog trainers, owners, and researchers have as much difficulty agreeing on a method for testing canine intelligence, as they do for human intelligence.

Certain breeds, like Border Collies, Poodles and Golden Retrievers, are generally easier to train than others. It is worth noting that these descriptions are relative to other dogs, not relative to the world at large. The ability to learn and obey commands is not the only possible measurement of intelligence.

Dogs are pack animals, meaning that they understand social structure and obligations and are capable of learning how to behave around other members of the pack. Adult canines train their young by correcting them when they behave in an unacceptable manner (biting too hard, eating out of turn, and so on) and reward them for acceptable behavior (by playing with them, feeding them, cleaning them, and so on). They are also den animals. This means that they can easily learn behavior related to keeping the den clean (such as housebreaking) and relaxing in an enclosed area (such as a crate during travel or for training).

Some breeds have been selectively bred for hundreds or thousands of years for the quality of learning quickly. That quality has been downplayed for other breeds in favor of other characteristics like the ability to track or hunt game, or to fight other animals. The capacity to learn basic obedience, however, and complicated behavior is inherent in all dogs. Owners must simply be more patient with some breeds than with others.

How Dog's Brain Work

Even though there is still a lot about dogs' brains that remain a mystery to us, we do know some interesting facts.

- Dog and human brains are structurally similar, but dogs have smaller brains;
- Dogs rely on smell much more than other pets, and 40% of their brain is used to analyze different scents;
- Dogs are known for their loyalty, and science has proved it - they prioritize their owner's scent over everything else;
- Dogs can count, understand up to 150 words and are thought to be as intelligent as 2-year-old children.

You see, it's not all treats and belly rubs. All dogs have active minds and they are wired to pick up on our cues, whether they're vocal or non-verbal.

Different types of Dog Intelligence

Dog intelligence comes in two basic varieties, says Today. There is "instinctive" intelligence, which is the type of intelligence and natural ability that is inherent to a breed or type of dog. This is the type of intelligence that makes terriers generally good at hunting small animals, bloodhounds good at finding things via scent, and border collies excellent at herding sheep.

The other type is "adoptive" intelligence, which is a dog's ability to learn. This category includes learning to perform certain tasks, socialization, and language comprehension. A dog may be stronger in one category of intelligence than the other. But just as a human who can rock a Mozart concerto on the piano is no better or worse than someone who can calculate the square root of pi without a calculator, a dog that is stronger in adoptive intelligence is no better or worse than one that shows more strength in the instinctive intelligence category. So, when conducting an IQ test on your dog, keep in mind that he may have natural talents and skills that the test doesn't take into account.

Dog IQ Test

Below are some tasks that you can do with your dog to test his IQ. These tasks are designed to evaluate a dog's ability to learn and retain information as well as his capacity for reasoning and problem solving. A scoring system is also provided so you can determine where your pooch falls on the dog IQ scale.

Task One: Place a large towel or blanket over your dog's head. This task will provide insight into your dog's ability to solve problems.

Scoring: Give your dog three points if he figures out how to free himself in under fifteen seconds, two points if it takes fifteen to thirty seconds, and one point if it takes more than thirty seconds.

Task Two: Place two or three empty buckets or cups upside down in a row. While your pup watches, place a treat under one of the containers. Distract him for a few seconds before allowing him to look for the treat. This test will help determine how well your dog can learn and retain information.

Scoring: Give him three points if he goes straight to the container hiding the treat, two if he checks one empty container before finding the right one, and one point if he checks both wrong containers before locating the treat.

Task Three: In a room where your pup has a favorite spot to lounge, send him out of the room and then rearrange the furniture. This task is designed to test his reasoning and cognition.

Scoring: Let your dog back into the room. If he goes right to his favorite spot, give him three points. Give him two points if he takes a short time to investigate before finding his spot. If he gives up and picks a new spot, give him one point.

Task Four: Underneath a piece of furniture (low enough to the ground that only your dog's paw will fit), place a treat within paw's reach. This task will test your dog's reasoning and problem-solving skills.

Scoring: Give your pooch three points if it takes him under a minute to reach for the treat using only his paw. If he tries to fit his head into the space first, or uses both his nose and paws, give him two points, and one point if he gives up entirely.

Task Five: At a time when you don't normally take your dog for a walk, pick up his leash while he's watching. This task tests your dog's ability to make and retain associations.

Scoring: Give your dog three points if he immediately gets the hint and becomes excited; two points if you need to walk to the door before he gets the message that it's time to go somewhere; and one point if he doesn't seem to understand what's going on.

Task Six: This one takes a little extra work on your part. With a piece of cardboard five feet wide and too tall for your dog to see over when standing on his back legs, build a barrier by attaching each end of the cardboard to a box big enough to support it. In the center of the cardboard, cut a rectangle about three inches wide that starts about four inches from the top and ends about four inches from the bottom. Toss a treat behind the barrier, allowing your dog to watch where it falls through the window you cut into the cardboard. This task will help measure reasoning and problem-solving ability.

Scoring: Give your dog three points if he takes 30 seconds (or less) to figure out he needs to walk around the barrier to get the treat. If he takes longer than 30 seconds to figure it out, give him two points, and one point if he tries to climb through the window or bulldoze his way through the barrier instead of going around.

- More than 15 points: Congratulations! Your dog is a genius.
- 13–15 points: Your dog's not Einstein but he's still a smart cookie.
- 9–12 points: Your dog won't be the class valedictorian but he'll get by okay.

- 5–8 points: Your dog might need some assistance figuring things out and getting things done.
- 1–4 points: It doesn't take brains to give cuddles and kisses—that's all that really matters, right?

Dogs that do well on these types of tasks tend to be highly trainable and make great candidates for service dogs. But again, these tests aren't infallible. Some dogs are simply stubborn, and their lack of cooperation probably has nothing to do with their intelligence. Some also say the smartest dogs are the ones that wait patiently for their owners to fork over a treat that they don't have to work for.

Perception

Perception refers to mental processes through which incoming sensory information is organized and interpreted in order to represent and understand the environment. Perception includes such processes as the selection of information through attention, the organization of sensory information through grouping, and the identification of events and objects. In the dog, olfactory information (the sense of smell) is particularly salient (compared with humans) but the dogs senses also include vision, hearing, taste, touch and proprioception. There is also evidence that dogs sense the earth's magnetic field.

One researcher has proposed that dogs perceive the passing of time through the dissipation of smells.

Awareness

The concept of "object permanence" refers to the ability of an animal to understand that objects continue to exist even when they have moved outside of their field of view. This ability is not present at birth, and developmental psychologist Jean Piaget described six stages in the development of object permanence in human infants. A similar approach has been used with dogs, and there is evidence that dogs go through similar stages and reach the advanced fifth stage by an age of 8 weeks. At this stage they can track "successive visible displacement" in which the experimenter moves the object behind multiple screens before leaving it behind the last one. It is unclear whether dogs reach Stage 6 of Piaget's object permanence development model.

A study in 2013 indicated that dogs appear to recognize other dogs regardless of breed, size, or shape, and distinguish them from other animals.

In 2014, a study using magnetic resonance imaging demonstrated that voice-response areas exist in the brains of dogs and that they show a response pattern in the anterior temporal voice areas that is similar to that in humans.

Social Cognition

Social Learning: Observation and Rank

An English Springer Spaniel taking cues from its master.

Dogs are capable of learning through simple reinforcement (e.g., classical or operant conditioning), but they also learn by watching humans and other dogs.

One study investigated whether dogs engaged in partnered play would adjust their behavior to the attention-state of their partner. The experimenters observed that play signals were only sent when the dog was holding the attention of its partner. If the partner was distracted, the dog instead engaged in attention-getting behavior before sending a play signal.

Puppies learn behaviors quickly by following examples set by experienced dogs. This form of intelligence is not particular to those tasks dogs have been bred to perform, but can be generalized to various abstract problems. For example, Dachshund puppies were set the problem of pulling a cart by tugging on an attached piece of ribbon in order to get a reward from inside the cart. Puppies that watched an experienced dog perform this task learned the task fifteen times faster than those left to solve the problem on their own.

The social rank of dogs affects their performance in social learning situations. In social groups with a clear hierarchy, dominant individuals are the more influential demonstrators and the knowledge transfer tends to be unidirectional, from higher rank to lower. In a problem-solving experiment, dominant dogs generally performed better than subordinates when they observed a human demonstrator's actions, a finding that reflects the dominance of the human in dog-human groups. Subordinate dogs learn best from the dominant dog that is adjacent in the hierarchy.

Following Human Cues

Dogs show human-like social cognition in various ways. For example, dogs can react appropriately to human body language such as gesturing and pointing, and they also understand human voice commands. For example, in one study, puppies were presented with a box, and shown that, when a handler pressed a lever, a ball would roll out of the box. The handler then allowed the puppy to play with the ball, making it an intrinsic reward. The pups were then allowed to interact with the box. Roughly three quarters of the puppies subsequently touched the lever, and over half successfully released the ball, compared to only 6% in a control group that did not watch the human manipulate the lever.

Similarly, dogs may be guided by cues indicating the direction of a human's attention. In one task a reward was hidden under one of two buckets. The experimenter then indicated the location of the reward by tapping the bucket, pointing to the bucket, nodding at the bucket, or simply looking at the bucket. The dogs followed these signals, performing better than chimpanzees, wolves, and human infants at this task; even puppies with limited exposure to humans performed well.

Dogs can follow the direction of pointing by humans. New Guinea singing dogs are a half-wild proto-dog endemic to the remote alpine regions of New Guinea and these can follow human pointing as can Australian dingoes. These both demonstrate an ability to read human gestures that arose early in domestication without human selection. Dogs and wolves have also been shown to follow more complex pointing made with body parts other than the human arm and hand (e.g. elbow, knee, foot). Dogs tend to follow hand/arm pointed directions more when combined with eye signaling as well. In general, dogs seems to use human cues as an indication on where to go and what to do. Overall, dogs appear to have several cognitive skills necessary to understand communication as information; however, findings on dogs' understanding of referentiality and others' mental states are controversial and it is not clear whether dog themselves communicate with informative motives.

For canines to perform well on traditional human-guided tasks (e.g. following the human point) both relevant lifetime experiences with humans—including socialization to humans during the critical phase for social development—and opportunities to associate human body parts with certain outcomes (such as food being provided by humans, a human throwing or kicking a ball, etc.) are required.

In 2016, a study of water rescue dogs that respond to words or gestures found that the dogs would respond to the gesture rather than the verbal command.

Memory

Episodic Memory

Dogs have demonstrated episodic-like memory by recalling past events that included the complex actions of humans.

Learning and using Words

Various studies have shown that dogs readily learn the names of objects and can retrieve an item from among many others when given its name. For example, in 2008, Betsy, a border collie, knew over 340 words by the retrieval test, and she was also able to connect an object with a photographic image of the object, despite having seen neither before. In another study, a dog watched as experimenters handed an object back and forth to each other while using the object's name in a sentence. The dog subsequently retrieved the item given its name.

In humans, "fast mapping" is the ability to form quick and rough hypotheses about the meaning of a new word after only a single exposure. In 2004, a study with Rico, a Border Collie, showed he was able to fast map. Rico initially knew the labels of over 200 items. He inferred the names of novel items by exclusion, that is, by knowing that the novel item was the one that he did not already know. Rico correctly retrieved such novel items immediately and four weeks after the initial exposure. Rico was also able to interpret phrases such as "fetch the sock" by its component words (rather than considering its utterance to be a single word). Rico could also give the sock to a specified person. This performance is comparable to that of 3-year-old humans.

In 2013, a study documented the learning and memory capabilities of a border collie, "Chaser", who had learned the names and could associate by verbal command over 1,000 words at the time of its publishing. Chaser was documented as capable of learning the names of new objects "by exclusion", and capable of linking nouns to verbs. It is argued that central to the understanding of the border collie's remarkable accomplishments is the dog's breeding background—collies bred for herding work are uniquely suited for intellectual tasks like word association which may require the dog to work "at a distance" from their human companions, and the study credits this dog's selective breeding in addition to rigorous training for her intellectual prowess.

Emotional Intelligence

Studies suggest that dogs feel complex emotions, like jealousy and anticipation. However, behavioral evidence of seemingly human emotions must be interpreted with care. For example, in his 1996 book *Good Natured*, ethologist Frans de Waal discusses an experiment on guilt and reprimands conducted on a female Siberian husky. The dog had the habit of shredding newspapers, and when her owner returned home to find the shredded papers and scold her she would act guilty. However, when the owner himself shredded the papers without the dog's knowledge, the dog "acted just as 'guilty' as when she herself had created the mess." De Waal concludes that the dog did not display true guilt as humans understand it, but rather simply the anticipation of reprimand.

One limitation in the study of emotions in non-human animals, is that they cannot verbalise to express their feelings. However, dogs' emotions can be studied indirectly through cognitive tests, called cognitive bias test, which measure a cognitive bias and allow to make inference about the mood of the animal. Researchers have found that dogs suffering from separation anxiety have a more negative cognitive bias, compared to dogs without separation anxiety. On the other hand, when dogs' separation anxiety is treated with medications and behaviour therapy, their cognitive bias becomes less negative than before treatment. Also administration of oxytocin, rather than a placebo, induces a more positive cognitive bias and positive expectation in dogs. It is therefore suggested that the cognitive bias test can be used to monitor positive emotional states and therefore welfare in dogs.

There is evidence that dogs can discriminate the emotional expressions of human faces. In addition, they seem to respond to faces in somewhat the same way as humans. For example, humans tend to gaze at the right side of a person's face, which may be related to the use of right brain hemisphere for facial recognition. Research indicates that dogs also fixate the right side of a human face, but not that of other dogs or other animals. Dogs are the only non-primate species known to do so.

Problem Solving

Sex-specific dynamics are an important contributor to individual differences in cognitive performance of pet dogs in repeated problem-solving tasks.

Captive-raised dingoes *(Canis dingo)* can outperform domestic dogs in non-social problem-solving. Another study indicated that after undergoing training to solve a simple manipulation task, dogs faced with an unsolvable version of the same problem look at the human, whereas socialized wolves do not. Modern domestic dogs use humans to solve their problems for them.

Learning by Inference

Dogs have been shown to learn by making inferences in a similar way to children.

Dogs have the ability to train themselves and learn behaviors through interacting and watching other dogs.

Theory of Mind

"Theory of mind" is the ability to attribute mental states—beliefs, intents, desires, pretending, knowledge, etc.—to oneself and others and to understand that others have beliefs, desires, intentions, and perspectives that are different from one's own. There is some evidence that dogs demonstrate a theory of mind by engaging in deception. For example, one observer reported that a dog hid a stolen treat by sitting on it until the rightful owner of the treat left the room. Although this could have been accidental, it suggests that the thief understood that the treat's owner would be unable to find the treat if it were out of view. A study found that dogs are able to discriminate an object that a human partner is looking for based on its relevance for the partner and they are more keen on indicating an object that is relevant to the partner compared to an irrelevant one; this suggests that dogs might have a rudimental version of some of the skills necessary for theory of mind.

Cat Intelligence

Cat intelligence is the considered capacity of learning, thinking, and problem solving ability possessed by the domestic cat. Cats learn by trial and error, observation and imitation. Cat learning abilities are aided by a good memory. Because of their sensitive sense of smell, cats prefer going outside to urinate and defecate.

Kittens need no training to use a litter box and cover up their waste; once they understand where the litter box is, they will seek it out from then on.

Cats can be trained to make use of a toilet.

Cats are able to form "learning sets," a skill once thought to be confined to primates. For example, cats that were trained to pull boxes on wheels showed they could combine that skill with their own insight to solve new problems. In one instance, a cat pulled the box to a specific location and used it, step-stool fashion, to reach a desired reward: a piece of food suspended from the ceiling by a string. The full extent of the cat's cognitive abilities is still largely unknown, but they continue to amaze their owners with their ingenious abilities, nevertheless.

Cats learn by observation, imitation, trial and error just as humans do. Stories abound which describe cats turning doorknobs to open doors, ringing doorbells, opening cupboards, turning off lights, and even using the toilet solely by observing the owner performing these activities. Many feline behaviorists as well as child psychologists seem to agree that the intelligence of an adult cat equals that of a 2 to 3 year old child.

While the cat is the most intelligent of all our domestic animals, it is important for cat owners to be clear about the limitation of their cat's thought processes, as attributing human motivations to the pet can interfere with the patient, methodical, approach necessary to treat behavior problems. For example, a cat cannot ponder the past or make plans for the future. It is therefore useless for an owner to punish a cat for something it did even a few minutes earlier or in the act, because the cat is incapable of making the connection between the action and the punishment. Conversely, its actions can not be motivated by revenge for some past insult by the owner. Stress caused by the owner's actions, and not spite, is the culprit here.

The Brain of a Cat

Brain Size

The brain of the domesticated cat is about 5 centimetres (2.0 inches) long, and weighs 25–30 g (0.88–1.06 oz). If a typical cat is taken to be 60 cm (24 in) long with a weight of 3.3 kg (7.3 lb), then the brain would be at 0.91% of its total body mass, compared to 2.33% of total body mass in the average human. Within the encephalization quotient proposed by Jerison in 1973, values above 1 are classified big brained, while values lower than 1 are small brained. The domestic cat is attributed a value of between 1–1.71 relative to human value that is 7.44–7.8. The largest brains in the cat kingdom are those of the tigers in Java and Bali, of which the largest relative brain size within the *pantera* is the *tigris*. It is debated whether there exists a causal relationship between brain size and intelligence in vertebrates. Correlations have been shown between these factors in a number of experiments. However, correlation does not imply causation. Most experiments involving the relevance of brain size to intelligence hinge on the assumption that complex behavior requires a complex (and therefore intelligent) brain; however, this connection has not been consistently demonstrated.

The surface area of a cat's cerebral cortex is approximately 83 cm^2 (13 in^2) whereas the human brain has a surface area of about 2,500 cm^2 (390 in^2). Furthermore, a theoretical cat weighing 2.5 kg (5.5 lb) has a cerebellum weighing 5.3 g (0.19 oz), 0.17% of the total weight.

Brain Structures

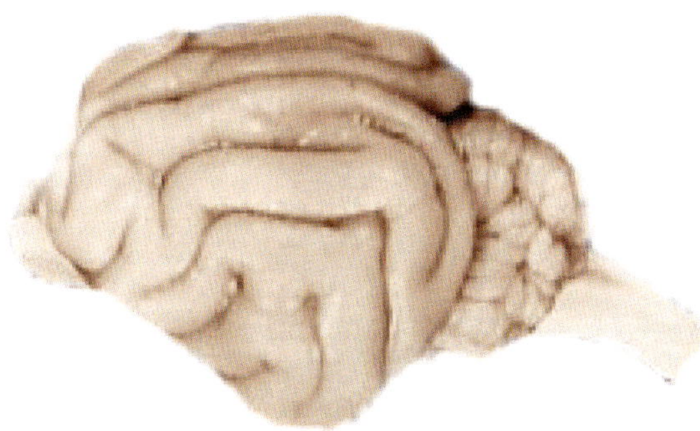

A cat's brain

According to researchers at Tufts University School of Veterinary Medicine, the physical structure of the brains of humans and cats is very similar. The human brain and the cat brain both have cerebral cortices with similar lobes.

The number of cortical neurons contained in the brain of the cat is reported to be 763 million. Area 17 of the visual cortex was found to contain about 51,400 neurons per mm^3. Area 17 is the primary visual cortex.

Both human and feline brains are gyrencephalic, i.e. they have a surface folding.

Analyses of cat brains have shown they are divided into many areas with specialized tasks that are extremely interconnected and share sensory information in a kind of hub-and-spoke network, with a large number of specialized hubs and many alternate paths between them. This exchange of sensory information allows the brain to construct a complex perception of the real world and to react to and manipulate its environment.

The thalamus of the cat includes a hypothalamus, an epithalamus, a lateral geniculate nucleus, and additional secondary nuclear structures.

Secondary Brain Structures

The domestic cat brain also contains the hippocampus, amygdala, frontal lobes (which comprise 3 to 3.5% of the total brain in cats compared to about 25% in humans), corpus callosum, anterior commissure, pineal gland, caudate nucleus, septal nuclei and midbrain.

Neuroplasticity

Grouse *et al.* 1979 ascertained the neuroplasticity of kittens' brains, with respect to control of visual stimulus correlated with changes in RNA structures. In a later study, it was found that cats possess visual-recognition memory, and have flexibility of cerebral encoding from visual information.

Brain and Diet

A cognitive support diet for felines is a food that is formulated to improve mental processes like attention, short and long-term memory, learning, and problem solving. Claims for cognitive support appear on a number of kitten formulations to help with brain development, as well as diets aimed at seniors to help prevent cognitive disorders. These diets typically focus on supplying Omega-3 fatty acids, omega-6 fatty acids, taurine, vitamins, and other supporting supplements that have positive effects on cognition.

The omega-3 fatty acids are a key nutrient in cognition for felines. They are essential for felines as they cannot be synthesized naturally and must be obtained from the diet. Omega-3 fatty acids that support brain development and function are alpha-linolenic acid, docosahexaenoic acid (DHA) and eicosapentaenoic acid (EPA). Fish oils, fish and other marine sources provide a very rich source of DHA and EPA. Alpha-linolenic acid can be acquired from oils and seeds.

Omega-6 fatty acids are also needed in feline cognition diets. The important omega-6 fatty acid that plays a role in brain support and cognition is arachidonic acid. Arachidonic acid or AA is found in animal sources such as meat and eggs. AA is required in cat diets, as felines convert insignificant amounts of it from linoleic acid due to the limited delta-6 desaturase. Like DHA, arachidonic acid is often found in the brain tissues of cats and seems to have a supporting role in brain function. In a 2000 study completed by Contreras *et al.*, it was found that DHA and AA made up 20% of the fatty acids in the mammalian brain. Arachidonic acid makes up high amounts in the membrane of most cells and has many pro-inflammatory actions.

Taurine is an amino acid, which is essential in cat diets due to their low capacity to synthesize it. Because taurine has the ability to cross the blood–brain barrier in the brain, it has been found to have a role in many neurological functions, especially in the visual development. Without taurine, felines can have an abnormal morphology in the cerebellum and visual cortex. When cats were fed a diet deficient in taurine, this leads to a decrease in the concentration of taurine in the retina of the eye. This results in deterioration of the photoreceptors, followed by complete blindness.

Choline is a water-soluble nutrient that prevents and improves epilepsy and cognitive disorders. Supplementation is part of therapy for cats with seizures and Feline cognitive dysfunction, despite this treatment being mostly based on anecdotal evidence and research done on dogs. It is the precursor to nerve chemicals like dopamine and acetylcholine, making it important for proper functioning of the nervous system.

Domestication Effects

Cat intelligence study is mostly from consideration of the domesticated cat. The process of domestication has allowed for closer observation of cat behaviour and in the increased incidence of interspecies communication, and the inherent plasticity of the cat's brain has become apparent as the number of studies in this have increased scientific insight. Changes in the genetic structure of a number of cats have been identified as a consequence of both domestication practises and the activity of breeding, so that the species has undergone genetic evolutionary change due to human selection (although this human selection has been coupled with an initial naturally occurring selective set of cats possessing characteristics desirable for the sharing of human habitation and living in Neolithic urban environments).

Cats' intelligence may have increased during their semi-domestication: urban living may have provided an enriched and stimulating environment requiring novel adaptive behaviours. This scavenging behaviour would only have produced slow changes in evolutionary terms, but such changes would have been comparable to the changes to the brain of early primitive hominids who co-existed with primitive cats (like, for example, Machairodontinae, Megantereon and Homotherium) and adapted to savannah conditions.

Cat's urban living is, however, unlikely to indefinitely improve the animal's intelligence: consider the fossil-based family tree of placental mammals above; the feline line diverged many years previously from the primate line; the cat both feral and domesticated is likely to be maintained in an evolutionary stasis by its niche position in the food web.

Exploitation Possibilities

Cats are known to be trained to perform as circus animals. An example of this is The Yuri Kuklachev Cat Theatre based in Moscow, the owner of which has been training cats for many years to perform a range of circus-style tricks.

In Artificial Intelligence

In November 2009, scientists claimed to simulate a cat's brain using a supercomputer containing 24,576 processors. This experiment did not simulate the function of the individual neurons in the brain, nor their synaptic patterns. It was intended to demonstrate that the problem of simulating a biological brain could be scaled to very large supercomputer platforms. However, the approach has been criticised as flawed.

There are a number of reasons for why the cat brain is a goal of computer simulations. Cats are familiar and easily kept animals, so the physiology of cats has been particularly well studied. The physical structures of human brains and cat brains are very similar. Cats, like humans, have binocular vision that gives them depth perception. Moreover, trying to build artificial mammal brains advances the research of both neuroscience and artificial intelligence.

Pigeon Intelligence

Pigeons are considered to be one of the most intelligent birds on the planet and able to undertake tasks previously thought to be the sole preserve of humans and primates. The pigeon has also been found to pass the 'mirror test' (being able to recognise its reflection in a mirror) and is one of only 6 species, and the only non-mammal, that has this ability. The pigeon can also recognise all 26 letters of the English language as well as being able to conceptualise. In scientific tests pigeons have been found to be able to differentiate between photographs and even differentiate between two different human beings in a photograph when rewarded with food for doing so.

Scientists recently announced that they have discovered that the human brain is laced with tiny magnetic particles made of magnetite. Magnetite is the same mineral that occurs naturally in lodestones. While the particles are distributed throughout the brain, they seem to be more concentrated in the membrane that encloses the brain. Scientists said that they had not seen the particles before because they are so tiny. The smallest are about a millionth of an inch in diameter while the largest are a hundred-thousandth of an inch in diameter. Added together, all the particles in the average brain amount to only one millionth of an ounce.

The particles are identical to those found in many bacteria, pigeons, salmon and whales. The particles help those creatures navigate using the earth's magnetic field. Scientists are not prepared to say that they serve the same purpose in humans. Instead, they continue to study the particles to learn what purpose they serve.

Based on evolutionary theory, scientists naturally conclude that the recognition of abstract forms such as letters of the alphabet requires the larger brain.

- Pigeons have the capacity to share attention between different dimensions of a stimulus, but (like humans and other animals) their performance with multiple dimensions is worse than with a single stimulus dimension.
- Pigeons can be taught relatively complex actions and response sequences, and can learn to make responses in different sequences.
- Pigeons readily learn to respond in the presence of one simple stimulus and withhold responding in the presence of a different stimulus, or to make different responses in the presence of different stimuli.
- Pigeons can discriminate between other individual pigeons, and can use the behaviour of another individual as a cue to tell them what response to make.
- Pigeons do less well with categories defined by abstract logical relationships, e.g. "symmetrical" or "same", though some experimenters have successfully trained pigeons to discriminate such categories.
- Pigeons seem to require more information than humans for constructing a three-dimensional image from a plane representation.
- Pigeons seem to have difficulty in dealing with problems involving classes of classes. Thus they do not do very well with the isolation of a relationship among variables, as against a representation of a set of exemplars.
- Pigeons can remember large numbers of individual images for a long time, e.g. hundreds of images for periods of several years.

All these are capacities that are likely to be found in most mammal and bird species. In addition pigeons have unusual, perhaps unique, abilities to learn routes back to their home from long distances. This homing behaviour is different from that of birds that learn migration routes, which usually occurs over a fixed route at fixed times of the year, whereas homing is more flexible; however similar mechanisms may be involved.

Pigeons showed mirror-related behaviours during the mirror test.

Discrimination Abilities of Pigeons

In a famous article in 1995, Watanabe, Sakamoto and Wakita described an experiment which showed that pigeons can be trained to discriminate between paintings by Picasso and by Monet. The birds were first trained on a limited set of paintings: when the shown painting was a Picasso, the pigeon was able to obtain food by repeated pecking; when it was a Monet, pecking had no effect. After a while, the pigeons would only peck when shown Picasso paintings. They were then able to generalize, and correctly discriminate between paintings of the two painters not previously shown, and even between cubist and impressionist paintings (cubism and impressionism being the two stylistic schools Picasso and Monet belong to). When the Monet paintings were shown upside down, the pigeons were not able to properly categorize anymore; showing the cubist works upside down did not have such an effect.

In 1995, the authors won the humorous Ig Nobel Prize in psychology for this work.

In a later paper, Watanabe showed that if pigeons and human college students undergo the same training, their performance in distinguishing between Van Gogh and Chagall paintings is comparable.

Similar experiments had shown earlier that pigeons can be trained to distinguish between photos showing human beings and those that do not, and between photos showing trees and those that do not, among many other examples.

In all these cases, discrimination is quite easy for humans, even though the classes are so complex that no simple distinguishing algorithm or rule can be specified. It has therefore been argued that pigeons are able to form "concepts" or "categories" similar to humans, but that interpretation is controversial. Nevertheless, the experiments remain important and often cited examples in cognitive science.

Theory of Mind in Animals

Whether any non-human animal can attribute mental states to others remains the subject of extensive debate. This despite the fact that several species have behaved as if they have a 'theory of mind' in various behavioral tasks. In this paper, we review the reasons of skeptics for their doubts: That existing experimental setups cannot distinguish between 'mind readers' and 'behavior readers', that results that seem to indicate 'theory of mind' may come from studies that are insufficiently controlled, and that our own intuitive biases may lead us to interpret behavior more 'cognitively' than is necessary. The merits of each claim and suggested solution are weighed. The conclusion is that while it is true that existing setups cannot conclusively demonstrate 'theory of mind' in non-human animals, focusing on this fact is unlikely to be productive. Instead, the more interesting question is how sophisticated their social reasoning can be, whether it is about 'unobservable inner experiences' or not. Therefore, it is important to address concerns about the setup and interpretation of specific experiments. To alleviate the impact of intuitive biases, various strategies have been proposed in the literature. These include a deeper understanding of associative learning, a better knowledge of the limited 'theory of mind' humans actually use, and thinking of animal cognition in an embodied, embedded way; that is, being aware that constraints outside of the brain, and outside of the body, may naturally predispose individuals to produce behavior that looks smart without requiring complex cognition. To enable this kind of thinking, a powerful methodological tool is advocated: Computational modeling, namely agent-based modeling and, particularly, cognitive modeling. By explicitly simulating the rules and representations that underlie animal performance on specific tasks, it becomes much easier to look past one's own biases and to see what cognitive processes might actually be occurring.

Several studies have been designed specifically to test whether animals possess theory of mind by using interspecific or intraspecific communication. Several taxa have been tested including primates, birds and canines. Positive results have been found; however, these are often qualified as showing only low-grade ToM, or rejected as not convincing by other researchers.

Methods

Specific categories of behaviour are sometimes used as evidence of animal ToM, including imitation, self-recognition, social relationships, deception, role-taking (empathy), perspective-taking, teaching and co-operation, however, this approach has been criticised. Some researchers focus on animals' understanding of intention, gaze, perspective, or knowledge, i.e. what another being has seen. Several experimental methods have been developed which are widely used or suggested as appropriate tests for nonhuman animals possessing ToM. Some studies look at communication between individuals of the same species (intraspecific) whereas others investigate behaviour between individuals of different species (interspecific).

Knower-Guesser

The Knower-Guesser method has been used in many studies relating to animal ToM. Animals are tested in a two-stage procedure. At the beginning of each trial in the first discrimination training stage, an animal is in a room with two humans. One human, designated the "Guesser," leaves the room, and the other, the "Knower," baits one of several containers. The containers are screened so that the animal can see who does the baiting, but not where the food has been placed. After baiting, the Guesser returns to the room, the screen is removed, and each human points directly at a container. The Knower points at the baited container, and the Guesser at one of the other three, chosen at random. The animal is allowed to search one container and to keep the food if it is found.

Competitive Feeding Paradigm

The competitive feeding paradigm approach is considered by some as evidence that animals have some understanding of the relationship between "seeing" and "knowing".

At the beginning of each trial in the paradigm, a subordinate animal (the individual thought to be doing the mind-reading) and a dominant animal are kept on opposite sides of a test arena which contains two visual barriers. In all trials, a researcher enters the enclosure and places food on the subordinate's side of one of the visual barriers (one baiting event), and in some trials the researcher re-enters the enclosure several seconds later and moves the food to the subordinate's side of the other visual barrier (second baiting event). The door to the subordinate's cage is open during any baiting by the researcher. The conditions vary according to whether the dominant's door is open or closed during the baiting events, and therefore whether the subordinate individual can see the dominant. After baiting, both of the animals are released into the test arena, with the subordinate being released several seconds before the dominant. If the animals possess ToM, it is expected that subordinates are more likely to gain the food, and more likely to approach the food under several circumstances: (1) When the dominant's door is closed during trials with a single baiting event; (2) when the dominant's door is open during a first baiting event but closed during a second; (3) in single baiting event trials with the dominant's door open, subordinates are more likely to get the food when they compete at the end of the trial with a dominant individual who did not see the baiting.

Goggles Method

In one suggested protocol, chimpanzees are given first-hand experience of wearing two mirrored visors. One of the visors is transparent whereas the other is not. The visors themselves are of markedly different colours or shapes. During the subsequent test session, the chimpanzees are given the opportunity to use their species-typical begging behaviour to request food from one of the two humans, one wearing the transparent visor and the other wearing the opaque. If chimpanzees possess ToM, it would be expected they would beg more often from the human wearing the transparent visor.

False Belief Test

A method used to test ToM in human children has been adapted for testing non-human animals. The basis of the test is to track the gaze of the animal. One human hides an object in view of a second human who then leaves the room. The object is then removed. The second human returns whereupon they will mistakenly look for the object where they last saw it. If the animal stares first and longest at the location where the human last saw the object, this suggests they expect him to believe it is still hidden in that place.

In Nonhuman Primates

Many ToM studies have used nonhuman primates (NHPs). One study that examined the understanding of intention in orangutans (*Pongo pygmaeus*), chimpanzees (*Pan troglodytes*) and children showed that all three species understood the difference between accidental and intentional acts.

Chimpanzees

There is controversy over the interpretation of evidence purporting to show ToM in chimpanzees.

William Field and Sue Savage-Rumbaugh have no doubt that bonobos have evolved ToM and cite their communications with a captive bonobo (*Pan paniscus*), Kanzi, as evidence.

However, empirical studies show that chimpanzees are unable to follow a human's gaze, and are unable to use other human-eye information. Attempts to use the "Goggles Method" on highly human-enculturated chimpanzees failed to demonstrate they possess ToM.

In contrast, chimpanzees use the gaze of other chimpanzees to gain information about whether food is accessible. Subordinate chimpanzees are able to use the knowledge state of dominant chimpanzees to determine which container has hidden food.

If chimpanzees can see two opaque boards on a table and are expecting to find food, they do not choose a board lying flat because if food was under there, it would not be lying flat. Rather, they choose a slanted board, presumably inferring that food underneath is causing the slant. Chimpanzees appear able to know that other chimpanzees in the same situation make a similar inference. In a foraging game, when their competitor had chosen before them, chimpanzees avoided the slanted board on the assumption that the competitor had already chosen it. In a similar study, chimps were provided with a preference box with two compartments, one containing a picture of food, the other containing a picture of nothing (the pictures had no causal relation to the contents). In a foraging competition game, chimpanzees avoided the chamber with the picture of food when their competitor had chosen one of the chambers before them. The authors suggested this was presumably on the assumption that the competitor shared their own preference for it and had already chosen it.

One study tested another sensory mode of ToM. In a food competition, a human sat inside a booth with one piece of food to their left and one to their right. The food could be withdrawn from the competing chimpanzee's reach when necessary. In the first experiment, the chimpanzee could approach either side of the booth unseen by the human, but then had to reach through either a transparent or opaque tube to get the food. In a second experiment, both were transparent and the human was looking away, but one of the tubes made a loud rattle when it was opened. Chimpanzees reached through the opaque tube in the first experiment and the silent tube in the second. The chimpanzees successfully concealed their food-stealing from their human competitor in both cases.

Chimpanzees have passed the False Belief Test involving anticipating the gaze of humans when objects have been removed.

Other Primates

Rhesus macaques selectively steal grapes from humans who are incapable of seeing the grape compared to humans who can see the grape.

In one approach testing monkeys, rhesus macaques (*Macaca mulatta*) are able to "steal" a contested grape from one of two human competitors. In six experiments, the macaques selectively stole the grape from a human who was incapable of seeing the grape, rather than from the human who was visually aware. The authors suggest that rhesus macaques possess an essential component of ToM: the ability to deduce what others perceive on the basis of where they are looking. Similarly, free ranging rhesus macaques preferentially choose to steal food items from locations where they can be less easily observed by humans, or where they will make less noise.

A comparative psychology approach tested six species of captive NHPs (three species of great apes: orangutans, gorillas, chimpanzees, and three species of old-world monkeys: lion-tailed macaques (*Macaca silenus*), rhesus macaques and collared mangabeys (*Cercocebus torquatus*)) in a "hide and seek" game in which the NHPs played against a human opponent. In each trial, the NHP has to infer where food has been hidden (either in their right or left hand) by the human opponent. In general, the NHPs failed the test (whereas humans did not), but surprisingly, performances between the NHP species did not reveal any inter-species differences. The authors also reported that at least one individual of each of the species showed (weak) evidence of ToM.

In a multi-species study, it was shown that chimpanzees, bonobos and orangutans passed the False Belief Test.

In 2009, a summary of the ToM research, particularly emphasising an extensive comparison of humans, chimpanzees and orang-utans, concluded that great apes do not exhibit understanding of human referential intentions expressed in communicative gestures, such as pointing.

In Birds

Parrots

Grey parrots (*Psittacus erithacus*) have demonstrated high levels of intelligence. Irene Pepperberg did experiments with these and her most accomplished parrot, Alex, demonstrated behaviour which seemed to manipulate the trainer, possibly indicating theory of mind.

Ravens

Ravens adjust their caching behaviour according to whether they have been watched and who was watching them.

Ravens are members of the corvidae family and are widely regarded as having complex cognitive abilities.

Food-storing ravens cache (hoard) their food and pilfer (steal) from other ravens' caches. They protect their caches from being pilfered by conspecifics using aggression, dominance and re-caching. Potential pilferers rarely approach caches until the storing birds have left the cache vicinity. When storers are experimentally prevented from leaving the vicinity of the cache, pilferers first search at places other than the cache sites. When ravens (*Corvus corax*) witness a conspecific making caches, to pilfer those caches they (1) delay approaching the cache only when in the presence of the storer, and (2) quickly engage in searching away from the caches when together with dominant storers. These behaviours raise the possibility that ravens are capable of withholding their intentions, and also providing false information to avoid provoking the storer's aggression to protect its cache. Ravens adjust their pilfering behaviour according to when the storers are likely to defend the caches. This supports the suggestion that they are deceptively manipulating the other's behaviour. Other studies indicate that ravens recall whom was watching them during caching, but also know the effects of visual barriers on what competitors can and can not see, and how this affects their pilfering.

Ravens have been tested for their understanding of "seeing" as a mental state in other ravens. It appears they take into account the visual access of other ravens, even when they cannot see the other raven.

In one study, ravens were tested in two rooms separated by a wooden wall. The wall had two functional windows that could be closed with covers; each cover had a peephole drilled into it. In the next familiarization step, the ravens are trained to use a peephole to observe and pilfer human-made caches in the adjacent room. Under test conditions, there was no other raven present in the adjacent room, however, a hidden loudspeaker played a series of sounds recorded from a competitor raven. The storing raven generalized from their own experience when using the peephole to pilfer the human-made caches and predicted that the audible (raven) competitors could potentially see their caches through the peep-hole and took appropriate action, i.e. the storing ravens finished their caches more quickly and they returned to improve their caches less often. The researchers pointed out that this represented "seeing" in a way that cannot be reduced to the tracking of gaze cues – a criticism leveled at many other studies of ToM. The researchers further suggested that their findings could be considered in terms of the "minimal" (as opposed to "full-blown") ToM recently suggested.

Using the Knower-Guesser approach, ravens observing a human hiding food are capable of predicting the behaviour of bystander ravens that had been visible at both, none or just one of two baiting events. The visual field of the competitors was manipulated independently of the view of the test-raven. The findings indicate that ravens not only remember whom they have seen at caching but they also take into account that the other raven's view was blocked.

Scrub Jays

Western scrub jays may show evidence of possessing theory of mind

Scrub jays are also corvids. Western scrub jays (*Aphelocoma californica*) both cache food and pilfer other scrub jays' caches. They use a range of tactics to minimise the possibility that their own caches will be pilfered. One of these tactics is to remember which individual scrub jay watched them during particular caching events and adjust their re-caching behaviour accordingly. One study with particularly interesting results found that only scrub jays which had themselves pilfered would re-cache when they had been observed making the initial cache. This has been interpreted as the re-caching bird projecting its own experiences of pilfering intent onto those of another potential pilferer, and taking appropriate action. Another tactic used by scrub jays is if they are observed caching, they re-cache their food when they are subsequently in private. In a computer modeling study using "virtual birds", it was suggested that re-caching is not motivated by a deliberate effort to protect specific caches from pilfering, but by a general motivation to simply cache more. This motivation is brought on by stress, which is affected by the presence and dominance of onlookers, and by unsuccessful recovery attempts.

In Dogs

Dogs can use the pointing behaviour of humans to determine the location of food.

Domestic dogs (*Canis familiaris*) show an impressive ability to use the behaviour of humans to find food and toys using behaviours such as pointing and gazing. The performance of dogs in these studies is superior to that of NHPs, however, some have stated categorically that dogs do not possess a human-like ToM.

The Guesser-Knower approach has been used with ToM studies in dogs. In one study, each of two toys was placed on the dog's side of two barriers, one opaque and one transparent. In experimental conditions, a human sat on the opposite side of the barriers, such that they could see only the toy behind the transparent barrier. The human then told the dog to 'Fetch' without indicating either toy in any way. In a control, the human sat on the opposite side but with their back turned so that they could see neither toy. In a second control, the human sat on the same side as the dog such that they could see both toys. When the toys were differentiable, dogs approached the toy behind the transparent barrier in experimental as compared to "back-turned" and "same-side" condition. Dogs did not differentiate between the two control conditions. The authors suggested that, even in the absence of overt behavioural cues, dogs are sensitive to others' visual access, even if that differs from their own. Similarly, dogs preferentially use the behaviour of the human Knower to indicate the location of food. This is unrelated to the sex or age of the dog. In another study, 14 of 15 dogs preferred the location indicated by the Knower on the first trial, whereas chimpanzees require approximately 100 trials to reliably exhibit the preference.

Human infants (10 months old) continue to search for hidden objects at their initial hiding place, even after observing them being hidden at another location. This perseverance of searching errors is at least partly contributed to by behavioural cues from the experimenter. Domestic dogs also commit more search errors in communicative trials than in non-communicative or non-social hiding trials. However, human-encultured wolves (*Canis lupus*) do not show this context-dependent perseverance in searching. This common sensitivity to human communication behaviour may arise from convergent evolution.

Dogs which have been forbidden to take food are more likely to steal the food if a human observer has their back turned or eyes closed than when the human is looking at them. Dogs are also more likely to beg for food from an observer whose eyes are visible compared to an observer whose eyes are covered by a blindfold.

In a study of the way that dogs interact, play signals were sent almost exclusively to forward-facing partners. In contrast, attention-getting behaviors were used most often when the other dog was facing away, and before signaling an interest to play. Furthermore, the type of attention-getting behaviour matched the inattentiveness of the playmate. Stronger attention-getting behaviours were used when a playmate was looking away or distracted, less forceful ones when the partner was facing forward or laterally.

In Pigs

An experiment at the University of Bristol found that one out of ten pigs was possibly able to understand what other pigs can see. That pig observed another pig which had view of a maze in which food was being hidden, and trailed that pig through the maze to the food. The other pigs involved in the experiment did not.

In Goats

A 2006 study found that goats exhibited intricate social behaviours indicative of high-level cognitive processes, particularly in competitive situations. The study included an experiment in which a subordinate animal was allowed to choose between food that a dominant animal could also see and food that it could not; those who were subject to aggressive behaviour selected the food that the dominant animal could not see, suggesting that they are able to perceive a threat based on being within the dominant animal's view – in other words, visual perspective taking.

Animal Navigation

True navigation is the ability of an animal to travel to a relatively precise target at a considerable distance without the need for familiar landmarks . To do this, the navigator must normally have a 'map' to show where it is relative to its goal. Having inferred the direction to the target from this map, the organism then needs a compass to steer itself along the appropriate vector. A great deal is known about the various redundant and self-calibrating compasses, used interchangeably or hierarchically as circumstances and the animal's experience dictate.

The thinking about animal maps has been largely shaped by work on pigeons that home to an imprinted loft location rather than migrating between winter and summer ranges. This makes homing pigeons a convenient model system, but it's important to keep in mind that they are homers rather than migrants; nor do they have a single map algorithm: homing requires flying experience near the loft and matures between about 6 weeks (fledging) and 12 weeks of age, during which time the gathering and processing of map information (and the use of compasses) changes dramatically. Researchers typically imagine that animals create, calibrate, and utilize the kind of two-dimensional graphic charts we call maps, but the juvenile salmon tested by Putman *et al.* (and, in retrospect, his hatchling sea turtles in earlier work) appear to have something like an innate look-up table which requires no prior experience or calibration.

Actually, the methods animals use to navigate their migration routes are even more amazing than an animal that could program a GPS device. Some of their navigation methods are so weird we don't really understand them.

The sun - This seems pretty simple. You can judge roughly what direction you're heading in by where the sun is. But factor in the time of day, time of year and cloud cover, and you're left with a pretty tricky navigation system. Yet starlings and ants navigate this way. Some birds can even travel at night using the sun -- theories suggest they take a "reading" from where the sun sets and use that to set their course. Others think that the polarization of light coming from the sun plays a role.

Wildebeest crossing creek bed in the Masai Mara, Kenya, Africa.

Landmarks - This is another pretty basic navigation system. Fly toward those mountains, head to the left a little when you see the ocean, and make a nest in the first nice-looking tree you can find. Whales traveling in the Pacific Ocean near the North American west coast use this method -- their landmark is hard to miss, because it's the entire continent of North America. They keep it on their left on the way south, and to their right when they head north.

Moon and stars - Planetarium experiments have proved that many birds rely on stellar cues to figure out which way to migrate. We can even tell which star they are orienting from.

Grant Faint

Scent - Once an animal is in the general area, scent can pinpoint specific locations. Scent won't get an animal from Saskatchewan to Mexico, but it probably helps salmon find their exact spawning ground, for instance. The scent of rain might shape wildebeest migrations.

Weather - Wind conditions are often used as supplementary navigation aid by birds. When deprived of other cues, such as the sun or stars, birds chose to fly downwind in an experiment. When the birds could see the sun and stars, they flew in the right direction regardless of wind direction.

Magnetic field - The earth has a magnetic field that's usually undetectable to humans who aren't holding a compass. Some animal species do have the ability to detect the magnetic field, however, and they use it to make their migrations. Bats and sea turtles use magnetic information to find their way. Some species of bacteria even rely on the magnetic field to orient themselves.

References

- Lea SEG (2010) Concept learning in nonprimate mammals: In search of evidence In D Mareschal, PC Quinn and SEG Lea, The Making of Human Concepts, pp. 173–199. Oxford University Press. ISBN 9780199549221
- Cazzolla Gatti, Roberto (2015). "Self-consciousness: beyond the looking-glass and what dogs found there". Ethology Ecology & Evolution. 28(2): 232–240. doi:10.1080/03949370.2015.1102777. ISSN 0394-9370
- Couchman, Justin J.; Coutinho, M. V. C.; Beran, M. J.; Smith, J. D. (2010). "Beyond Stimulus Cues and Reinforcement Signals: A New Approach to Animal Metacognition" (PDF). Journal of Comparative Psychology. 124 (4): , 356 –368. doi:10.1037/a0020129. PMC 2991470. PMID 20836592
- Gopnik, Alison (May 2016). "How Animals Think; A new look at what humans can learn from nonhuman minds". The Atlantic. Retrieved 25 April 2016
- O'Connell, Caitlin (2007). The Elephant's Secret Sense: The Hidden Lives of the Wild Herds of Africa. New York City: Simon & Schuster. pp. 174, 184. ISBN 0-7432-8441-0
- Hof, P. R.; Van der Gucht, E. (Jan 2007). "Structure of the cerebral cortex of the humpback

whale, Megaptera novaeangliae (Cetacea, Mysticeti, Balaenopteridae)". Anat Rec. 290 (1): 1–31. doi:10.1002/ar.20407. PMID 17441195

- Child, Ben (3 September 2014). "The elephant who cried: Raju's story on way to big screen". The Guardian. London: Guardian News and Media. Retrieved October 19, 2014
- Nissani,, M., Hoefler-Nissani, D., Lay, U.T.& Htun, U.W. (2005). "Simultaneous visual discrimination in Asian elephants". Journal of the Experimental Analysis of Behavior. 83: 15–29. doi:10.1901/jeab.2005.34-04. PMC 1193698
- Linden, Eugene (2002). The Octopus and the Orangutan: More Tales of Animal Intrigue, Intelligence and Ingenuity. New York City: Plume. pp. 16–17, 104–105, 191. ISBN 0-452-28411-2. OCLC 49627740
- Horowitz, Alexandra (2009). "Attention to attention in domestic dog (Canis familiaris) dyadic play". Journal Animal Cognition. Springer Berlin / Heidelberg. 12 (1): 107–18. doi:10.1007/s10071-008-0175-y. PMID 18679727
- K Guo; C Hall; S Hall; K Meints; D Mills (2007). "Left gaze bias in human infants, rhesus monkeys, and domestic dogs". Perception. 36 ECVP. Retrieved 24 June 2010
- Corsin A. Müller1, Kira Schmitt, Anjuli L.A. Barber, Ludwig Huber (2015). "Dogs can discriminate the emotional expressions of human faces". Current Biology. 25 (5): 601–605. doi:10.1016/j.cub.2014.12.055. PMID 25683806

Permissions

All chapters in this book are published with permission under the Creative Commons Attribution Share Alike License or equivalent. Every chapter published in this book has been scrutinized by our experts. Their significance has been extensively debated. The topics covered herein carry significant information for a comprehensive understanding. They may even be implemented as practical applications or may be referred to as a beginning point for further studies.

We would like to thank the editorial team for lending their expertise to make the book truly unique. They have played a crucial role in the development of this book. Without their invaluable contributions this book wouldn't have been possible. They have made vital efforts to compile up to date information on the varied aspects of this subject to make this book a valuable addition to the collection of many professionals and students.

This book was conceptualized with the vision of imparting up-to-date and integrated information in this field. To ensure the same, a matchless editorial board was set up. Every individual on the board went through rigorous rounds of assessment to prove their worth. After which they invested a large part of their time researching and compiling the most relevant data for our readers.

The editorial board has been involved in producing this book since its inception. They have spent rigorous hours researching and exploring the diverse topics which have resulted in the successful publishing of this book. They have passed on their knowledge of decades through this book. To expedite this challenging task, the publisher supported the team at every step. A small team of assistant editors was also appointed to further simplify the editing procedure and attain best results for the readers.

Apart from the editorial board, the designing team has also invested a significant amount of their time in understanding the subject and creating the most relevant covers. They scrutinized every image to scout for the most suitable representation of the subject and create an appropriate cover for the book.

The publishing team has been an ardent support to the editorial, designing and production team. Their endless efforts to recruit the best for this project, has resulted in the accomplishment of this book. They are a veteran in the field of academics and their pool of knowledge is as vast as their experience in printing. Their expertise and guidance has proved useful at every step. Their uncompromising quality standards have made this book an exceptional effort. Their encouragement from time to time has been an inspiration for everyone.

The publisher and the editorial board hope that this book will prove to be a valuable piece of knowledge for students, practitioners and scholars across the globe.

Index